Gas and LNG Price Arbitrations

A Practical Handbook, Second Edition

Consulting Editors **James Freeman** and **Mark Levy QC**

Consulting editors
James Freeman and Mark Levy QC

Managing director
Sian O'Neill

Gas and LNG Price Arbitrations, Second Edition
is published by

Globe Law and Business Ltd
3 Mylor Close
Horsell
Woking
Surrey GU21 4DD
United Kingdom
Tel: +44 20 3745 4770
www.globelawandbusiness.com

Gas and LNG Price Arbitrations, Second Edition

ISBN 9781787421929
EPUB ISBN 9781787421936
Adobe PDF ISBN 9781787421943
Mobi ISBN 9781787421950

Table of contents

Drafting an effective price review clause

Mark Levy QC
Allen & Overy LLP

The other chapters in this book identify a range of issues that can arise both before and during the course of a price review arbitration. This chapter draws together the various strands and considers why those issues arise, the extent to which they are often the product of imperfect drafting and how the uncertainties that these issues create might be avoided.

1. The nature of pricing disputes

The focus of this book is on gas price review disputes. Disputes are, by their nature, unpredictable. Usually, at the outset of the parties' contractual relationship, no one knows whether a dispute will arise, when it will arise or what it will be about. However, price review clauses are rare creatures in the context of commercial agreements, in that they provide a structure which actively encourages parties to identify, on a periodic basis, a dispute which can then be referred to arbitration.

Different laws place different emphasis on the primacy of the contractual language. However, whatever the governing law, the ultimate purpose of any arbitral tribunal should be to apply the bargain to the dispute. Even if the parties' entire bargain is not to be found in the agreement itself, the terms of the agreement will invariably be highly determinative to the outcome. How the price review clause is drafted will therefore usually play a significant part in shaping how a tribunal resolves any price review dispute.

There is growing user dissatisfaction with arbitral awards in this area. The most common complaint appears to be that tribunals arrive at decisions which are divorced from the commercial reality of what those in the industry could possibly have intended or wished for. The most frequently cited example is the *Atlantic LNG* decision, where the tribunal appeared to reach a decision that neither party sought or wanted.[1] Specifically, the tribunal amended the price formula in a way that neither party had proposed. (*Atlantic LNG* became public only because it was challenged before US federal courts.) While there is some

1 *Atlantic LNG Company of Trinidad & Tobago v Gas Natural Aprovisionamientos* SDG SA, UNCITRAL, Final Award dated 17 January 2008.

sympathy with this view, the *Atlantic LNG* tribunal could 'rewrite' the sale and purchase agreement (SPA) only because the price review clause afforded the tribunal wide discretion. In fact, in its decision confirming the *Atlantic LNG* award, the US federal court in New York expressly noted that, in the absence of any restrictions in the price review clause, the arbitral tribunal could amend the formula the way it wanted to.[2] Thus, if the parties clearly tell a tribunal what it must do and how it must do it, the tribunal is compelled to do as it is told. Tribunals meander off course because they are given scope to do so in the agreement.

There are various ways in which the parties can prevent tribunals from reaching unpredictable results. They can add strict parameters which, for example, limit the extent to which the tribunal can amend the price formula. Just as importantly, however, parties should pay attention to the drafting of the clause and provide as clear guidance to the tribunal as possible. It is remarkable how many disputes in this area, involving hundreds of millions or even billions of dollars, arise out of, or even because of, vague and ambiguous drafting. It is true that once a dispute arises, ambiguity can be a lawyer's friend. It is also the case that parties sometimes consciously choose to leave open how a provision is to be applied. As a general rule, however, commercial parties going into a dispute crave certainty as to the result. Effective drafting can go a considerable way towards providing such certainty.

Part of the problem is that many current long-term agreements contain price review clauses which belong to a bygone era. Price review clauses first emerged before the days of gas-to-gas competition, when European gas was commonly priced by reference to a basket of competing alternative fuels. However, while the pricing dynamics in markets have changed, the contractual wording in many price review clauses has not kept pace. There is remarkable homogeneity in the drafting and current clauses still bear a lot of resemblance to those clauses that commentators such as Brautaset were commenting on 15 years ago.[3] To the extent that experienced industry negotiators are relying on an assumption that everyone in the industry understands what these terms mean, this is dangerous. Absent compelling evidence, there can be no confidence that a tribunal will respect that assumption, even if it is factually correct.

The days when complex European gas pricing disputes were determined by panels steeped in industry knowledge, through an informal and non-adversarial procedure, have largely disappeared. Today, these disputes are usually presided over by experienced commercial arbitrators. These tribunals are adept at considering contested complex factual evidence and applying those facts to the

---

2 *Atlantic LNG Company of Trinidad & Tobago v Gas Natural Aprovisionamientos* SDG, SA, 2008 WL 4344525 (SDNY, 16 September 2008).

3 A Brautaset, E Høiby, RO Pedersen and CF Michelet, *Norsk Gassavsetning, Rettslige Hovedelementer (Norwegian Gas Sales, Principal Juridical Foundations)* (Sjørettsfondet, Oslo, 1998).

law, but will often have very little direct, or even indirect, experience of gas pricing. While there are some experienced gas pricing specialists who sit as arbitrators, they are few and far between. It is rare to find one pricing specialist on a panel, let alone two or three. The effect of this is that general commercial arbitrators are often being asked to grapple with concepts with which they have no familiarity or experience. That can lead to unpredictable results. That problem is exacerbated if they are making this decision within the context of ambiguous drafting.

The rest of this chapter considers some common drafting issues and provides thoughts on what steps parties might take at the contractual, and pre-contractual, stage either to avoid disputes arising or to strengthen their position in case disputes do arise.

2. Pre-contractual negotiations

Long-term gas agreements often follow lengthy pre-contractual negotiations. During these negotiations, there will invariably be at least a discussion regarding the terms of the price review clause. When a dispute emerges as to what the parties intend the clause to mean, it is not uncommon for one party to rely on these discussions to support the construction that it urges the tribunal to adopt.

The extent to which pre-contractual negotiations are admissible will be a matter of the governing law. For example, under English law, such evidence is generally considered inadmissible as to interpretation. A contract under English law is to be interpreted objectively, in order to ascertain the meaning which the document would convey to a reasonable person having all of the knowledge which would reasonably have been available to the parties at the time of the contract.[4] The determination is therefore conducted by taking into account evidence as to the factual background to the contract, provided that such evidence was reasonably available to the parties. However, it is an established principle that evidence of the previous negotiations of the parties and their declarations of subjective intent is inadmissible.[5]

However, even under English law, the window is not entirely closed to using such evidence. It may be used to support a claim of rectification or estoppel by convention.[6] Evidence of pre-contractual negotiations may also be admissible where the communications between the parties, including their pre-contractual negotiations, demonstrate that they habitually used words in an unconventional sense. This 'private dictionary' principle applies only where the meaning alleged is unconventional and cannot operate to choose between

4 *Investors Compensation Scheme Ltd v West Bromwich Building Society* [1998] 1 All ER 98 at 114.
5 *ibid*, at 114–15.
6 *Chartbrook Ltd v Persimmon Homes Ltd* [2009] UKHL 38 at 42.

conventional alternatives.[7] Although this principle is narrow in scope, one can see how it might apply in the context of a price review clause, where parties sometimes adopt language which has a particular meaning to them, based on usage in the gas business, which differs from its conventional meaning.

Other laws are less restrictive as to the extent to which evidence of pre-contractual negotiations might be used. For example, under Dutch law, contract terms are to be interpreted by determining the meaning that the parties in the given circumstances could reasonably have given to them and what they could reasonably have expected from each other in that respect.[8] This determination is conducted by taking into account all relevant circumstances of the case,[9] including the pre-contractual negotiations.[10] Correspondence exchanged between the parties during the negotiation of the agreement (emails, drafts of the agreement and minutes of the negotiation meetings), as well as witness testimony on what was discussed during the negotiations, can be relevant to the interpretation of a contract term and as such will be admissible evidence in a dispute about the construction of a contract term.

Similarly, under French law, contractual construction is closely related to the concept of the common intention of the parties. Where the wording of the contract is clear and precise, every contractual term must be applied strictly. However, where the wording of the contract is ambiguous, the judge should use the rules of construction set out in Articles 1156 to 1164 of the Civil Code.[11] According to Article 1156 of the code, "one must in agreements seek what the common intention of the contracting parties was, rather than pay attention to the literal meaning of the terms". In order to establish the common intention of the parties, French courts can take into account "external factors" such as the parties' behaviour before and after they entered into the contract, pre-contractual negotiations (especially when set out in pre-contractual documents),[12] letters exchanged between the parties before they entered into the contract[13] or previous agreements between the parties.[14] Pre-contractual negotiations can therefore play an important role in contractual construction.[15]

The position under New York law is somewhere between the two extremes. When a contract is unambiguous, a court or tribunal cannot consider extrinsic evidence (including evidence of pre-contractual negotiations) to interpret the contract terms. However, if a court or tribunal determines that specific contract

7 *ibid*, at 45.
8 Dutch Supreme Court, 13 March 1981, NJ 1981/635 (Haviltex).
9 Dutch Supreme Court, 20 February 2004, NJ 2005/493 (DSM/Fox).
10 Dutch Supreme Court, 13 January 1978, NJ 1978/302 (Swimming Pool/Diligentia) and Dutch Supreme Court, July 1 1982, NJ 1983/682 (Codicil).
11 P Malaurie, L Aynès and P Stoffel-Munck, *Les obligations, Defrénois*, 3rd edn (Paris) section 772.
12 French Court of Cassation, Civ 1, 18 February 1986, Bull Civ I, No 31.
13 French Court of Cassation, Com, 8 December 1987, Bull Civ IV, No 262.
14 French Court of Cassation, Soc, 21 February 1957, Bull Civ IV, No 205.
15 F Terré, P Simle and Y Lequette, *Les obligations*, 10th edn (Dalloz, Paris) section 189 in fine and section 450.

language is ambiguous, it can consider evidence of the parties' negotiations, minutes from meetings, prior drafts of the agreement or fact witness testimony to explain the ambiguous contract term.[16] If a court or tribunal determines that a term of a contract is ambiguous, the parties may offer extrinsic evidence to explain the ambiguous term to show the objective intent of the parties.[17]

The extent to which pre-contractual negotiations may be admissible thus varies markedly, depending on the governing law. However, as can be seen above, even in jurisdictions where pre-contractual negotiations are not generally admissible, the door is often not entirely closed. Further, international arbitration tribunals tend to apply rules of evidence less strictly than courts. It is therefore not unheard of for a party to adduce evidence of pre-contractual negotiations before the tribunal if it believes that such evidence supports its case, even if the strict laws of evidence do not permit it. While a party on the receiving end of such evidence may wish to apply to the tribunal for an order that it be struck out as inadmissible, such orders are relatively rare (as the tribunal can direct itself to ignore the evidence), and a party may be reluctant to draw even more attention to the evidence by making such an application. Indeed, even if such an application succeeds, the damage may already be done in terms of the impression made before the tribunal. Far better, therefore, at least to have the option of presenting cogent evidence which contradicts that which has been adduced (even if it is presented on the basis that it is without prejudice to the position that such evidence should be given no weight).

Given the relatively high chance that a price review clause will be the subject of a dispute, and that different versions will be put forward by the parties in any dispute as to what their intentions were, thought should be given to this issue by in-house counsel at the negotiating stage. At the very least, counsel should ensure that someone on the negotiating team keeps a full and legible[18] record of discussions regarding the clause and its negotiation history (including full meeting notes). It may also be prudent to produce an agreed form of minutes, although in the frenzy of a deal completion, this may not be realistic.

3. Common drafting ambiguities

The same interpretational issues arise again and again in price review disputes. As noted above, part of this is due to the fact that the same imperfect clauses seem to be reused as standard in long-term supply contracts. It is also often the case that a phrase or concepts whose meaning may seem 'obvious' to parties at

16 *Chapman v NY State Div for Youth*, 546 F 3d 230, 236 (2d Cir 2008).
17 *WWF Paper Corp v Quinlan* 00 Civ 0512 (LAK), 2001 US Dist. LEXIS 3444, at *3 (SDNY 29 March 2001).
18 Relying on handwritten notes can cause problems. By the time that the dispute arises, the author of the note may well have left the company, meaning that it may be very difficult to locate his or her notes and decipher them, even if they can be found.

the time of the commercial discussions can be anything but to a panel of arbitrators, particularly when faced with diametrically opposed accounts from counsel as to what was intended.

The following is a list – by no means exhaustive – of words or phrases which are often central to pricing disputes. Despite their importance, many of these phrases are capable of meaning different things to different parties and different tribunals. It is worth considering at the drafting stage what different arguments can be raised in respect of what these clauses mean, and whether it is sensible to close down these ambiguities by specific drafting. As noted above, the parties might consciously decide to leave these issues deliberately open, so as to leave open the scope for maximum flexibility (and argument) down the line. But in those instances, there needs to be a proper understanding of what the implications might be.

3.1 Trigger event

Although there is no 'standard' price review clause, one common feature of price review mechanisms is their periodic application. Because long-term sale and purchase agreements (SPAs) can last for upwards of 20 years, they often allow either party to serve a price review notice every so many years (often three years). In addition, it is common for long-term SPAs to provide either party with a fixed number of 'wildcard' reviews, which can be used outside of the usual cycle and in special circumstances.

Irrespective of the number of times that the price review mechanism can be applied, European long-term contracts have traditionally required that to reopen the price, the requesting party must first show that a trigger event has occurred. The definition of a 'trigger event' can vary markedly across price review clauses. At one end of the spectrum is the *Sonatrach-Distrigas SPA*, which provides that the parties will meet "every four years" to discuss revisions to the contract price. In other words, the trigger event occurs automatically with the passage of time and neither party need show anything more to establish entitlement to a price review.

At the other end of the spectrum are those price review clauses which require the requesting party to show that certain qualitative changes have taken place in the relevant market (eg, changes in economic circumstances) before the price can be reopened. The *Atlantic LNG* SPA is a good example in this regard. That SPA provided that the requesting party must show that

> *economic circumstances in Spain beyond the control of the Parties, while exercising due diligence, have substantially changed as compared to what it reasonably expected when entering into this Contract ... and the Contract Price ... does not reflect the value of Natural Gas in the Buyer's end-user market.*

The wording of the *Atlantic LNG* SPA, while by no means standard, is quite typical of the wording found in other SPAs in the Atlantic Basin. Despite the similarities, it is not uncommon to see different arbitral tribunals faced with different SPAs reaching different conclusions in respect of the same events, such as liberalisation in European gas markets or decrease in demand for gas in the power sector. Often these differences in tribunals' conclusions can be due to the differences in the trigger criteria.

Some interpretive issues that often arise in respect of such clauses are as follows:

- 'Significant changes in circumstances' – are the parties precluded from relying on certain changes, such as changes which are foreseeable?
- 'Beyond the control of the parties' – changes within the control of the parties are often expressly excluded as triggering events. However, does this mean that a party cannot rely on a trigger event if the other party has caused the event, or simply that a change must be beyond the control of the party that wishes to rely on it? This particular ambiguity can easily be resolved by clear drafting.
- Some clauses require an adjustment to 'take into account' or 'reflect' the change in value caused by the change in circumstances. Questions can arise as to whether this requires the tribunal to isolate the change in value caused by the change in circumstances and then reflect it in the revised formula. Evidentially, this can be a challenge. If 'taking into account' is meant to mean something less specific than actually transposing a particular change in value into an adjustment, what precisely does it mean?

One way in which parties can avoid these interpretive difficulties, while at the same time not conferring an automatic right to reopen the price, is to provide mechanical trigger criteria in their SPA. For example, the SPA in *Esso Exploration & Production UK Limited v Electricity Supply Board (Esso v ESB)* effectively required the contract price to deviate from the market price by a fixed percentage before the price could be reviewed.[19] The advantage of such clauses is that they provide greater certainty to the parties in respect of circumstances under which the price review mechanism will be triggered, and therefore reduce the possibility of uncertain outcomes. However, the parties in *Esso v ESB* were able to draft their price review clause in the way that they did because the relevant market, the UK, has a liquid trading hub (the National Balancing Point), which market participants accept reflects the market price of gas in the UK. This is by no means true in every market, although it is becoming increasingly true of Western Europe. Over time, one is likely to see price review clauses across Europe providing for more mechanical and objective trigger criteria as in *Esso v ESB*.

19 *Esso Exploration & Production UK Limited v Electricity Supply Board* [2004] EWHC 723 (Comm).

Another option is to dispense with the trigger requirement altogether. This is a common feature of many new Asian long-term contracts, which often simply provide for a periodic review of the price without any need to show a change in circumstances. The basis for the adjustment is then prescribed by what is usually quite narrow and objective criteria. While this tends to make periodic adjustments more inevitable, it does at least limit the scope for disputes.

To the extent there is a requirement to satisfy the definition of a 'trigger event', parties should also be careful about how they refer to the trigger events in their price review notices. Price review clauses typically require the party seeking a price review to serve a notice (commonly known as the 'price review notice') setting out the trigger event (or events) which it believes have given rise to a right to a price review. Once a price review notice has been validly served, the parties are usually required to enter into negotiations for a fixed period, at the end of which, if the parties have failed to reach an agreement, the dispute can be submitted to arbitration. One issue that can arise in the arbitration is whether the trigger events on which the requesting party is relying in the arbitration were properly identified in the price review notices. If not, it might be argued that the party's claims in respect of the unidentified trigger events are inadmissible and/or the tribunal lacks jurisdiction over those claims.

The tribunal in International Chamber of Commerce (ICC) Case 9812 had to consider a very similar issue.[20] The claimant's price review request specified the "impact of the forthcoming gas tax" on the price of natural gas as the only relevant economic change for seeking a price review.[21] However, in its notice of arbitration, the claimant mentioned several other circumstances, including changes in the border value of processed gas, changes in the movements in oil prices relative to oil prices in neighbouring countries, and the decline in the premium of natural gas.[22] In its award, the tribunal distinguished between:

- new changes which do fall within the scope of the original price review request and therefore cannot be considered in the arbitration; and
- "additional or different substantiation in support of the [price review] request" which form part of the arbitration.[23]

The tribunal also explained that any substantiation of a price review request which is not "specifically mentioned" in the original price review request must be:

- "within the scope of the request for price review"; and
- "sufficiently connected to the identified circumstance(s)".[24]

20 ICC Case 9812 (ICC International Court of Arbitration Bulletin, Vol 20/2, 2009) Final Award, p74.
21 *ibid.*
22 *ibid.*
23 *ibid.*
24 *ibid.*

The changes alleged by the claimant, except for the introduction of a gas tax, were neither mentioned in the original price review request nor connected to the gas tax. As such, the tribunal held that those changes could not be considered in the arbitration.[25]

A similar issue has also been considered by the English courts. In *Esso v ESB*, the Commercial Court had to consider whether an arbitral tribunal had jurisdiction to determine a gas price review dispute in circumstances where the seller had allegedly served an invalid price review request.[26] The SPA in *Esso v ESB* provided that either party could serve a price review request only when, among other things, it was "reasonably satisfied" in good faith that the contract price was below a certain comparator.[27] The court concluded that because the seller had committed an error in calculating the relevant comparator, its price review request was invalid. Consequently, no dispute had arisen that was capable of being referred to arbitration and the arbitrators lacked jurisdiction to determine the dispute.[28]

3.2 Price adjustment

As for trigger clauses, the extent to which a tribunal can revise the price formula varies significantly. As noted above, an immediate issue that a tribunal is likely to face is whether it should revise the price formula only to the extent necessary to reflect the proven change in circumstances, or whether it can effectively rewrite the formula. Both scenarios can prove challenging. In the former case, it may be difficult to isolate the effect of the proven change on the value of gas from other changes. Moreover, once that effect has been isolated, it may be difficult to transpose accurately into the revised price formula. In the latter case – that is, when the tribunal has the discretion to rewrite the price formula – it must consider what, if any, guidance the parties have provided in their contract. As noted above, in the absence of sufficient guidance, the tribunal can, like the *Atlantic LNG* tribunal, reach a conclusion that neither party requested.

Some interpretive issues that commonly arise in the context of price adjustment are outlined below.

(a) 'In any case' clause

As discussed in the 'trigger phase' chapter, this is a common clause which takes its name from its standard opening words. The clause usually reads something like: "[Whatever adjustment is made] in any case the buyer should be entitled to market the gas economically." The question therefore arises as to whether this gives the buyer an independent right to market the gas economically, effectively giving rise to a self-standing ground for an adjustment in itself.

---

25 *ibid*, p74.
26 [2004] EWHC 723 (Comm), para 2.
27 *ibid*, para 6.
28 *ibid*, para 27.

Even if this is not the case, a further dispute can arise as to whether the clause operates only a 'shield' – that is, it applies only in the case of an upward adjustment, acting as a ceiling on any increase that the seller might otherwise be entitled to – or whether it can also act as a 'sword' to increase the price reduction that a buyer might otherwise achieve to a level that enables it to sell the gas competitively.

There are arguments for and against all of these positions. Different tribunals will arrive at different positions, depending on the precise terms of the clause and the relevant circumstances of the case. However, this all gives rise to significant uncertainty. It is hard to see the benefit of deliberate ambiguity here – far better for the parties to address these uncertainties in the original drafting, rather than leave it in the hands of the tribunal to determine.

(b) ***'Market economically'***

What does this mean? Does it mean, for example, that the buyer should always be entitled to make a net profit on the gas which it sells so that the seller is effectively taking the price risk under the contract and guaranteeing the buyer a profit? If so, at what level should the profit be fixed? Is it fixed over the life time of the contract? Is the buyer entitled to sell the gas economically into all segments of the market – even the least profitable ones, where other sellers are losing money – or should it be aggregated across all segments?

(c) ***'Value'***

Common issues that arise include the following:

- Is value the same as price?
- Where is the value to be assessed – at the border, at the end-user level or across the entire spectrum of the market?
- When is value to be assessed? Is it at the date of delivery under a contract which may have been negotiated many months prior to delivery? Or is it at the date at which parties agree to buy and sell the gas, although delivery may be outside the review period (in which case, how is the actual price to be assessed)?

4. Other guidance which the parties might want to give the tribunal

As noted above, the scope for tribunals to stray off course is proportional to the extent to which the language of the provision permits them. Generally speaking, the more prescriptive the clause, the more predictable the outcome should be. Of course, there is always a balance between aiming for certainty and crafting a clause that has sufficient flexibility to deal with unexpected changes in circumstances that may occur over the course of a long-term agreement. In addition to ironing out drafting ambiguities, thought should be given as to how prescriptive the parties want to be in directing the tribunal as to how it should make its decision. In this author's view, this is highly desirable. It takes discretion out of the tribunal's hands and puts it back in the hands of the parties. The objection to over-prescription is that the clause needs to be sufficiently flexible to deal with the uncertainties that may arise over the course of a 25-year agreement. This is true. However, the two aims are not mutually preclusive. One can direct the tribunal as to how it should (or should not) arrive at its decision without depriving it of the flexibility that it needs to deal with changes in the market that are unforeseen at the time that the agreement was concluded. To the extent that there is concern that over-prescription may make the clause unworkable, adding wording such as "so far as practicable" will often ensure that the machinery continues to function in all scenarios.

Again, the list below is far from exhaustive. It sets out some examples where guidance might be given to tribunals:

- Comparators – if a tribunal is being asked to compare the value of gas in a market to the contract price, do the parties want to specify what the tribunal should be making the comparison with, so as to ensure an 'apples to apples' comparison? For example, should the tribunal compare the contract price only to the price of gas in 'comparable contracts' – a phrase sometimes used in these provisions – although, again, what does 'comparable' mean? Should a tribunal considering a price review under, say, a long-term liquified natural gas (LNG) contract consider only the prices under other long-term LNG contracts, even if the supplies under that contract also compete with pipeline gas? How close must the contracts which are being cited as comparators be in duration to that under review? This issue is linked to what is said below about evidence of prices. The risk of providing a strict category of comparators (eg, other LNG contracts) is that there may not be much publicly available information from which to draw a representative body of evidence.
- Most price clauses require a comparison between two periods of time (at the beginning of the price review period and then at the end of the period) in order to assess whether there has been a change in circumstance and also, sometimes, whether there has been a change in value. On the face of it, this should be straightforward. However, it is not instructive to compare the value of gas on a day in the middle of summer to that in the middle of winter. Providing the tribunal with guidance on how to deal with seasonality can save a lot of argument.
- As noted above, a dispute may arise as to whether value at a certain date should be assessed by reference to delivered prices at that date (which may reflect prices negotiated some time previously) or agreements struck during the relevant period (even if the gas is to be delivered outside the relevant period and assuming that the prices at which the agreements

have been struck for future delivery are known or can be reliably ascertained).

- Prices delivered into the end-user market reflect many 'non-gas' costs, such as transportation, regasification and margins. There may be an argument as to how best to strip out these costs to ascertain the true market value of gas. Two of the most commonly used techniques are an assessment of border prices (so that the value of gas can be considered before the non-gas costs are added), or a so-called 'netback', where the costs are deducted from the end-user prices to determine the cost of gas at the border. Arguments can arise with regard to both approaches. A party may question the reliability of customs prices, which are often published by national authorities.[29] Similarly, to carry out a netback, one must know the various constituent elements, including the end-user prices, various costs and margins. However, there may not be publicly available and/or authoritative sources from which information can be gathered on these various elements. Moreover, while one might think that the value of gas that one calculates following either approach should be similar, for a variety of reasons this is often not the case. The value differential can be substantial, and so can give rise to significant disagreements between the parties and their respective experts. Thought should therefore be given as to whether the parties wish to provide guidance to the tribunal as to what is the preferred approach.

5. Evidence of prices

The issue of the price of gas unsurprisingly often features heavily in these disputes. This can be a contentious topic in itself. The problem is often exacerbated by a lack of certainty as to how one establishes what the price of gas is. In many markets, publicly available evidence of prices in certain sectors (eg, the power sector, where the gas is bought to produce electricity, rather than being sold directly to end-users) can be hard to come by – which leads to two questions:

- Is it satisfactory to rely on publicly available data, even if it is not necessarily reliable? Some jurisdictions capture prices in customs data. This has the advantages of being publicly available and having been put together objectively. However, arguments are sometimes put forward that this data has limitations which can undermine its veracity.
- To what extent is it permissible for parties to rely on their own prices in the market to establish value? It might be argued that these can be a more reliable source than publicly available data; however, such pricing

29 The validity of the objection will of course depend on the specifics of the data in question and the terms of the relevant contract.

information will usually be available only to the buyer which on-sells to end-users in its home market. This can lead to an asymmetrical situation where the buyer can either choose to rely on its own prices (if they are helpful) or not disclose them if it does not consider them helpful, while at the same time criticising the publicly available sources – which are the only sources of information on which the seller can rely.

These problems may be resolved if the buyer is required to disclose its prices. However, some clauses limit expressly the scope of disclosure in price review disputes and it may be hard to compel disclosure if the buyer chooses not to rely on that information itself. In any event, disclosure is sometimes resisted in respect of commercially sensitive information such as prices. Although confidentiality in itself may not be a ground for refusing disclosure, it might influence the tribunal's decision. Furthermore, some tribunals take the view that in international arbitration, it is for each party to make its case and it is not permissible for a party to seek disclosure from the other party in order to substantiate its own case; rather, information can be sought only to test the other party's case. This can give rise to the potential asymmetry of information as described above.

These are difficult questions and it is unrealistic to expect them all to be resolved in the drafting of the clause, however desirable this might be. Nevertheless, thought should be given at the time that the clause is drafted as to whether:

- the parties do want to specify whether the tribunal should be entitled to base its decision on publicly available data; and/or
- a party which holds confidential pricing information should be required to disclose it.

6. Conclusion

While price review arbitrations share many characteristics of commercial arbitrations, they are in many respects quite different. They require the tribunal to invade a space that is normally the preserve of the parties – the negotiated price. This requires a commercial perception that is beyond the experience, if not the reach, of many tribunals. This can lead to disastrous and unexpected results. To some extent, that is an inescapable function of the process. However, the level of risk can be reduced by clear drafting and by giving the tribunal careful guidance.

The trigger phase

Marnix Leijten
Martje Verhoeven-de Vries Lentsch
De Brauw Blackstone Westbroek NV

1. Introduction

In this chapter, we examine the trigger phase and discuss the criteria for a price review, identifying issues arising in arbitration where the tribunal must determine whether these criteria have been met. The chapter is based on our experience with long-term agreements related to the sale of natural gas and LNG, of which many different forms and many different price review clauses exist. Specific contracts can and will differ in many important respects. We aim to provide a general overview of the criteria for a price review we have encountered in these agreements.

Generally, the core of the price review clause is relatively short. Guidance for both parties and the arbitral tribunal on its construction is limited because there are only a few publically available awards.[1] Most, but not all, legal systems we have come across consider evidence from the negotiation history in one way or another when determining the proper construction of the clause. Yet, very often the negotiation history – which is in most cases ambiguous and regularly dates back to the 1980s and 1990s – shows that the wording of the price review clause is based on commercial compromises, making it difficult to identify the parties' intentions and the underlying principles. In addition, certain concepts – such as gas-to-gas competition and the development of spot markets – were not always foreseen at the time the price review clause was negotiated. Therefore, in practice, establishing the criteria for a price review can be difficult for both parties and tribunals, which may struggle to interpret and apply the price review clause correctly. Since any determination in this respect by a tribunal will have important effects for the parties in future price reviews, and since the financial

1 There are two well-known International Chamber of Commerce (ICC) awards, in 1999 and 2007, regarding the same long-term gas sales agreements. They were published in 2009 in the ICC *Bulletin*, "Extracts from ICC Arbitral Awards: Price Setting and Price Revision in the Energy Sector" (2009) 20(2) *ICC International Court of Arbitration Bulletin* 69–76 and 93–109. In addition, an LNG price review between Gas Natural and Atlantic LNG conducted in New York City in accordance with the UNCITRAL Arbitration Rules became public through the case of *Gas Natural Aprovisionamientos, SDG, SA v Atlantic LNG Company of Trinidad and Tobago* in the United States District Court for the Southern District of New York (2008 WL 4344525 (SDNY)). Gas Natural filed a petition for confirmation of the award, while Atlantic LNG opposed the petition and filed a motion to vacate the award. While the award itself was not published, the price review clause and the price adjustment ordered by the tribunal were.

interests at stake are often significant only in one review, it is crucial to get the interpretation right. It is often difficult to determine the meaning of one specific review clause in a specific contract between specific parties. Making general statements regarding the criteria for a price review is even more so, but that is the aim of this chapter.

It has been established, and is commonly accepted, that the main rationale of the price review clause is to provide both parties with a tool to restore for the future the economic balance of the contract as it was agreed to in the past. If this economic balance becomes distorted, the price review clause serves to reset it by adjusting the price provisions.[2] As a result, in price review arbitrations the main focus is on expert opinions regarding the economic (im)balance of the contract and the adjustments to the price provisions that may follow therefrom.

Taking the above into account, in this chapter we discuss the main elements of the price review clause and the criteria that typically are derived from it and which must be met in order successfully to request an adjustment of the price provisions. We focus on the relevance of the review period and the various elements of the requirement of a change in the economic circumstances, as well as the buyer's ability economically to market the gas. However, first we discuss a more procedural element of the price review clause: the purpose of the trigger letter and its validity.

2. Validity of the trigger letter

A price review generally occurs only at the request of one or both of the parties. Under any long-term gas sales agreement, the price review clause contains one or more provisions on the formalities of requesting a price review. Generally, the contract provides that this request should be made in writing. In practice, the party requesting a price review does so in a letter to the other party. This letter is referred to as the trigger letter. Although the letter becomes less relevant once the parties are in arbitration, its validity is crucial to start a price review and to set the scene for and determine the scope of the requested review.

2.1 When to send a trigger letter

The price review clause usually contains several provisions on the formalities of the request for a price review. Typically, it provides that each of the parties is entitled to request a price review with effect at a certain date. This date is not only the date on which the price review comes into effect (ie, the day that the adjusted price provisions – if any – become effective), but also determines the deadline for sending a request for a price review, which generally should be made before such trigger date.[3] In general, the price review clause is drafted in

---

2 Final Award in Case 13504, January 2007, "Extracts from ICC Arbitral Awards: Price Setting and Price Revision in the Energy Sector" (2009) 20(2) *ICC International Court of Arbitration Bulletin* 94.

3 Final Award in Case 9812, August 1999, "Extracts from ICC Arbitral Awards: Price Setting and Price Revision in the Energy Sector" (2009) 20(2) *ICC International Court of Arbitration Bulletin* 73.

such a way that it provides the relevant trigger date for the entire duration of the contract by prescribing a certain set interval of one or multiple years between each price review.[4]

This is illustrated in Example 1.

Example 1

A price review clause may provide that each of the parties can first request a price review on 1 October 2012 and then on 1 October every following three years – that is, on 1 October 2015, 1 October 2018 and so on until the end of the contract. In each case the party that wishes to request a price review will have to send its trigger letter to the other party before the trigger date, for instance, before 1 October 2021.

A request for a price review sent after the relevant trigger date is likely to be considered invalid.

In addition to regular price reviews, some contracts also provide for a limited number of review options within the prescribed intervals, also known as joker or wildcard price reviews.[5] The contract will stipulate how many of these reviews can be requested by each party during the term of the contract. Normally, they can be requested at any time. The only requirement is that the requesting party must send the trigger letter on or before the date on which it wishes the price review to come into effect.

Thus, in the above example a wildcard price review could be triggered on 1 January 2020, which is between the 1 October 2018 and 1 October 2021 trigger dates. The party triggering a wildcard will have to make a careful assessment and balance its interests to determine when exactly to do so: the later it triggers the wildcard, the clearer the distortion of the economic balance may have become and thus the case for a review will be stronger. However, the disadvantage of waiting is that, usually, the price adjustment will apply only from the date of the trigger.

2.2 Level of specificity

What should be included in a trigger letter? Usually, the contract provisions are fairly broad. Typically, apart from a reference to the contract and the date on which the price review comes into effect, the contract provides that a request for a price review should also substantiate the reasons for the request. However, the contract rarely provides the level of detail needed to meet this requirement. Based on the more general requirement of substantiation of the request, it is

4 See also Stephen P Anway and George M von Mehren, "The Evolution of Natural Gas Price Review Arbitrations" in *The Guide to Energy Arbitrations – Second Edition* (June 2017), available at https://globalarbitrationreview.com/chapter/1142622/the-evolution-of-natural-gas-price-review-arbitrations.

5 *ibid.*

common practice for a party requesting a price review to specify which significant changes in the economic circumstances it relies on when requesting a price review.[6]

Generally, the description of such changes will be broad to avoid narrowing the scope of the price review too early in the process. As the trigger letter is sent before the effective date and thus some information that is crucial to make a proper analysis is not yet available, parties take into account that further analyses may lead to different conclusions after the trigger letter has been sent.

Depending on the contractual language, the trigger letter is generally considered to limit the scope of the price review to the topics addressed therein. For instance, if the trigger letter focuses on developments in one market, the price review should be limited to that market and cannot subsequently be expanded in arbitration. A party that has included a very specific and detailed description of the changes in its trigger letter but later deviates from this may therefore face a formal defence from its counterparty regarding the scope of the price review. Therefore, typically, the requesting party includes more general market developments justifying an adjustment of the price provisions. Usually, the requesting party does not substantiate its request with relevant data and more in-depth market analyses until the first meetings in the early stages of the negotiations. We do not believe that it would be easy to deem a trigger letter invalid because it is not specific enough. The general purpose of the letter is to inform the other party and, since negotiations are to follow, a high level of detail is not necessary at this stage. However, problems could arise if the triggering party seeks to add completely new review grounds during the course of the negotiations or in arbitration that are not covered by the trigger letter. It should be possible for the parties to conduct and conclude a review relatively quickly after the trigger letter. Adding review grounds weeks or months after the letter, based on new facts, is inconsistent with that goal.[7]

2.3 Claim to be included?

Should the requesting party's actual claim – the actual adjustment of the price provisions – be included in the trigger letter? Defendant parties have argued this point in arbitrations. Normally, the contract does not explicitly provide for this requirement and so it could be assumed that it is not required. However, it could be argued that substantiating the request would also involve the actual adjustment

6 Final Award in Case 9812, August 1999, "Extracts from ICC Arbitral Awards: Price Setting and Price Revision in the Energy Sector" (2009) 20(2) *ICC International Court of Arbitration Bulletin* 73.

7 *ibid*, p73: "[I]t must be assumed that the price request may include other not specifically mentioned circumstances, provided they are within the scope of the request for price review and sufficiently connected to the identified circumstance(s) … a party failing to provide convincing arguments cannot be precluded from initiating arbitral proceedings and submitting additional or different 'substantiation' in support of the request, provided that any such further substantiation is within the scope of the price review request."

that the requesting party seeks. Most parties have already spent a considerable amount of time analysing the changes and determining the requested adjustment before requesting a price review. However, as noted above, it is likely that not all relevant data is available when the trigger letter is sent. This makes it difficult for the requesting party to specify its claim in its trigger letter. The purpose of the trigger letter is to inform the other party that a review is requested and of the general circumstances on which the request is based, so precise details of the changes sought should not be necessary (again, depending on the precise wording of the clause). Clarity on this can and should be given in the negotiations that follow, without much prejudice for the party that receives the trigger letter.

Parties have found a way around the lack of essential data at the date of sending the trigger letter: they can agree to maintain the official trigger date and to enter into commercial negotiations instead of sending a formal trigger letter. If the negotiations fail, the trigger letter can still be sent. Based on the arrangement made, this trigger would then be deemed to be valid as of the official trigger date. This method is applied regularly and allows time not only for commercial negotiations, but also for further analyses of the changes and the required adjustments. Usually, this enables the requesting party to provide a specification or at least an indication of the adjustment of the price provisions that it seeks before the official trigger letter is sent. If a party has prepared an indication of the adjustment that it requires before sending the trigger letter, one could argue that the requesting party should in that case inform the other party of such adjustment (and thus its claim) in the trigger letter. However, price review clauses do not usually explicitly prescribe this, it is not an established practice, and we have not seen any arbitral tribunals apply a sanction because the requested adjustment has not been included in the trigger letter (indeed, sometimes a party does not even specify its claimed adjustment in its request for arbitration).

Parties typically also agree on an extension of the deadline to submit a trigger letter when a previous price review is not concluded yet because of a pending arbitration. This makes sense, as it will be virtually impossible to determine whether one actually wants to trigger a price review and on which basis, let alone its claim, if it is unclear what are the current price provisions. The extension to send a trigger letter for the next price review applies to both parties. Again, the effective date does not change, meaning the trigger letter may be sent long thereafter. Although this provides a limited advantage to the party triggering the price review, because more data will generally have become available, it is primarily a practical solution that is necessary to deal with the uncertainty of the outcome of the pending arbitration.

3. Criteria for a price review

As stated above, the main purpose of a price review is to restore the balance of the long-term gas sales agreement if it has become distorted. Although the wording of the price review clause provides reference to the criteria that have to be met in order for a party to claim a price review leading to an adjustment of the price provisions, these criteria are not as straightforward as they seem.[8] In this section, we discuss the various criteria and the issues that can be identified when determining whether the criteria have been met.

3.1 Changes in the economic circumstances

A typical price review clause first of all requires that the economic circumstances have changed compared with what is reflected in the price provisions of the contract. Thus, the parties should identify:

- the changes that have occurred in the economic circumstances; and
- to what extent these changes are not reflected in the price provisions.[9]

Generally, there is no limit to the type of changes in the economic circumstances that a party can rely on in a price review.[10] They can vary from general economic or market conditions to specific legislation. Commonly, the price review clause contains an explicit requirement that the change must have some impact on the energy market as a whole or on the value of gas specifically, as a change in the economic circumstances which can be reflected in the adjusted price provisions. As most parties requesting an adjustment of the price provisions ultimately rely on a change in the value of gas, this change is discussed separately in section 3.2.

Although there is generally no limit to the type of changes in the economic circumstances, there are usually some further requirements that have to be met. A party will have to prove that the change in the economic circumstances:

- has taken place within the review period;
- is significant;
- has occurred in the country or market of the buyer (or another relevant market defined in the contract); and
- is outside of the control of the parties.[11]

---

8 An example of a typical price review clause can be found in the two awards rendered in 1999 and 2007, "Extracts from ICC Arbitral Awards: Price Setting and Price Revision in the Energy Sector" (2009) 20(2) *ICC International Court of Arbitration Bulletin* 69–70 and 94–95.

9 A similar approach was taken in the two published awards. Final Award in Case 9812, August 1999 and Final Award in Case 13504, January 2007, "Extracts from ICC Arbitral Awards: Price Setting and Price Revision in the Energy Sector" (2009) 20(2) *ICC International Court of Arbitration Bulletin* 70–71 and 96.

10 Guy Block, "Arbitration and Changes in Energy Prices: A Review of ICC Awards with respect to Force Majeure, Indexation, Adaptation, Hardship and Take-or-Pay Clauses" (2009) 20(2) *ICC International Court of Arbitration Bulletin* 59–60.

11 Similar conditions are mentioned in the Final Award in Case 9812, August 1999, "Extracts from ICC Arbitral Awards: Price Setting and Price Revision in the Energy Sector" (2009) 20(2) *ICC International Court of Arbitration Bulletin* 70–71. These are also referred to by Block, n 10 above, p59. See also Marco Lorefice, "Gas Price Review Arbitrations: Changing the Indexation Formula" in *The Guide to Energy Arbitrations – Second Edition* (June 2017), available at https://globalarbitrationreview.com/chapter/1142625/gas-price-review-arbitrations-changing-the-indexation-formula#_ftnref8.

In addition, price review clauses can also include the additional requirement that the changes in economic circumstances are enduring or long-lasting in order to avoid parties triggering and obtaining a price review based on a temporary anomaly.

Although these requirements may seem straightforward, many parties and tribunals struggle with how to interpret and apply these four requirements, and determine whether these are actually met.[12] Each can be considered to refer to a separate element, which is discussed below.

(a) *Time element – changes within the review period*

In determining whether there has been a change in the economic circumstances, the point at which the change occurred must be established by comparing the position at time A and time B. This is determined by the so-called 'review period'.

As indicated above, the price review clause will stipulate when a party can request a price review and when such price review comes into effect. Thus, it also sets the period to which the price review relates. This period, also referred to as the 'review period', runs from the date when the previous price review could have been triggered until the date when the current price review can be triggered.

In Example 1 given in section 2.1: if a party requests a price review with effect as of 1 October 2021, the review period runs from 1 October 2018 until 1 October 2021.

This also applies when a party has triggered a wildcard price review. In that case, the end of the review period is set by the effective date of the wildcard review, whereas the beginning of the review period is determined by the effective date of the previous price review.

In Example 1: when a wildcard price review is triggered on 1 January 2020, the review period runs from 1 October 2018 to 1 January 2020. If triggered in a timely fashion, the subsequent review period will run from 1 January 2020 until 1 October 2021.

Thus, the review period plays an important role in a price review, as it defines the period in which the changes in the economic circumstances should have occurred. Traditionally the general approach in this regard was very strict: the change had to have occurred within the review period.[13] However, there is some flexibility in this area, in particular since the financial crisis began in 2008 and the European spot markets have developed into liquid gas markets.[14]

---

12 See also Mark Levy, "Gas Price Review Arbitrations: Certain Distinctive Characteristics" in *The Guide to Energy Arbitrations – Second Edition* (June 2017), available at https://globalarbitrationreview.com/chapter/1142623/gas-price-review-arbitrations-certain-distinctive-characteristics.

13 See Final Award in Case 9812, August 1999 and Final Award in Case 13504, January 2007, "Extracts from ICC Arbitral Awards: Price Setting and Price Revision in the Energy Sector" (2009) 20(2) *ICC International Court of Arbitration Bulletin* 70 and 97.

Today, much of the gas is sold on a forward basis against forward prices for delivery in the future, which can be as much as several gas years[15] later. As a consequence, there have been many discussions on when – if at all – a change in such forward values of gas can be considered to fall within the review period. Is the date of delivery and payment for the gas the relevant date for consideration, the date of entering into the forward sale or any date or period in between, and against which gas price under the parties' contract is this forward sale compared? Even though a change may occur outside of the review period, it can still be relied on if its effects occurred within the relevant review period. Before, a typical example would be new legislation which has been announced to enter into force at a later date. If it comes into effect after the review period, technically this change would fall outside the review period. However, as market participants usually start adjusting their behaviour in anticipation of such new legislation before it will enter into force, this adjustment in behaviour may qualify as a change in the economic circumstances within the review period if it affects the value of the gas. However, the risk of a change falling between two periods should at all times be prevented. Similarly, a party cannot rely on the same change twice, unless it can show that the change has not been reflected in the price provisions.

The relevance of the review period demonstrates why a party may opt for a wildcard price review instead of a regular price review. Typically, this is done when the severity of the impact of the changes in the economic circumstances does not permit the requesting party to wait for the next price review. Alternatively, a party may take a more opportunistic approach by trying to 'lock in' a sudden and perhaps temporary change in the market, thus targeting the ideal moment to request a wildcard price review.

When combining the time element with the required change in economic circumstances, the requesting party should prove that these changes occurred within the review period, and that they are not yet reflected in the price provisions. This is determined by comparing the situation that relates to the change at the start and end of the review period. The observed change should be compared with the change in the contract price (determined by comparing the contract price at the start and end of the review period) to verify the extent to which the change is reflected in the contract price.

---

14 These developments sparked a wave of price review arbitrations. See Stephen P Anway and George M von Mehren, "The Evolution of Natural Gas Price Review Arbitrations" in *The Guide to Energy Arbitrations – Second Edition* (June 2017), available at https://globalarbitrationreview.com/chapter/1142622/the-evolution-of-natural-gas-price-review-arbitrations and Marco Lorefice, "Gas Price Review Arbitrations: Changing the Indexation Formula" in *The Guide to Energy Arbitrations – Second Edition* (June 2017), available at https://globalarbitrationreview.com/chapter/1142625/gas-price-review-arbitrations-changing-the-indexation-formula#_ftnref8.

15 Gas years run from 1 October until 30 September the following year.

(b) *Weight element – significant changes*

Typically, the price review clause also provides that the economic circumstances should change significantly compared with what is reflected in the price provisions. The term 'significant' seems to serve as a minimal threshold, ensuring that the change has sufficient weight to justify an adjustment of the price provisions.[16]

When can a change be judged to be significant? The price review clause does not usually contain any further quantification of this criterion, and leaves the parties and the tribunal with little guidance in determining whether this requirement has been met. The reference to the price provisions suggest that the significance is related to the change in the value of gas. But what change in value is significant? A reduction of 0.01 eurocent per kilowatt hour may not seem significant, but as under these long? term gas sales agreements extremely large volumes are being sold, this seemingly small change is definitely significant.

What is clear is that the requirement of significance does not intend effectively to tum the price review clause into a hardship clause.[17] However, it forces the tribunal to consider whether the nature and the extent of the change justify a price review and adjustment of the price provisions. This must be assessed on a case-by-case basis.

(c) *Influence element – changes beyond the control of the parties*

In order to prevent misuse of the price review clause, generally it will provide that the changes in the economic circumstances should be beyond the control of the parties (or, at least, the party requesting the review).[18] Thus, the price review clause requires that both parties did not and were not in a position to influence the changes that they invoke in a price review. However, in principle this should not stop the requesting party from relying on changes that were (allegedly) caused by the other party.

This raises questions regarding state-owned entities and parties subject to regulatory action. When can they be considered to be in control of certain changes? In the case of monopolists or dominant market players that have price-setting power, can they be considered responsible for certain developments in their market? Although these points are known to have been raised and have merit in some cases, today's extensive competition laws and regulations are known to have created more liberal gas markets, where no single

---

16 See also Final Award in Case 9812, August 1999, "Extracts from ICC Arbitral Awards: Price Setting and Price Revision in the Energy Sector" (2009) 20(2) *ICC International Court of Arbitration Bulletin* 71.

17 See also Peter Roberts, *Gas Sales and Gas Transportation Agreements, Principles and Practice*, 3rd edn (Sweet & Maxwell, 2011) 15-005, p189.

18 See also Final Award in Case 9812, August 1999 and Final Award in Case 13504, January 2007, "Extracts from ICC Arbitral Awards: Price Setting and Price Revision in the Energy Sector" (2009) 20(2) *ICC International Court of Arbitration Bulletin* 71 and 98.

player has the ability to control changes. Therefore, typically tribunals will find that the changes are beyond the parties' control. However, the situation may be different if a party shows atypical market behaviour which is related to the change that it invokes in the price review.

(d) Geographical element – changes in the country or the market of the buyer

Typically, the price review clause also includes a geographical element, providing that the change must have occurred in the buyer's country or market, or another defined relevant market or market area. Not all price review clauses define this geographical element, although some do refer to a certain country or region in which the buyer traditionally operates.

The question remains how to interpret this geographical limitation. In general, this interpretation is relatively strict. When the price review clause refers to a specific country, the main focus lies on the changes that have occurred in that country.[19] If there is no clear definition of the geographical area, parties tend to focus on the changes that occurred in the country in which the buyer sells the majority of its gas.[20]

Although this assessment may have been relatively straightforward in the past, it raises a number of questions in today's gas market. Borders are fading as gas companies have expanded their business across country lines. In addition, local gas markets are merging. Finally, over the years spot markets have emerged and expanded across Europe, increasing in liquidity. These developments raise various questions regarding the definition of the buyer's country and market, and create challenges for parties and arbitrators. Depending on the review period and the wording of the price review clause, this may lead to extensive discussions.

By extension, it is also relevant whether the trigger letter referred to (developments in) just one market while the buyer is active in other markets as well. In that case, it can be argued that any adjustments made to the price provisions should be limited to reflect only the (developments in the) market mentioned in the trigger letter, and that potential other (developments in other) markets should not be taken into account.

The developments in the market of the buyer also trigger yet another, but related, discussion regarding the market level that should be taken into account in price reviews when determining the value of gas. This is examined in section 3.2.

---

19 See also Final Award in Case 13504, January 2007, "Extracts from ICC Arbitral Awards: Price Setting and Price Revision in the Energy Sector" (2009) 20(2) *ICC International Court of Arbitration Bulletin* 97.

20 In the Final Award in Case 13504, the tribunal considered that, although the buyer was reselling the gas only into part of the country of the buyer, "documentation of economic changes in [buyers' state] as a whole would suffice to meet the requirement for a price review". Final Award in Case 13504, January 2007, "Extracts from ICC Arbitral Awards: Price Setting and Price Revision in the Energy Sector" (2009) 20(2) *ICC International Court of Arbitration Bulletin* 97.

3.2 Changes in the value of gas

As discussed above, in principle a party may rely on any type of change in the economic circumstances, provided that the requirements discussed above are met. However, most price review clauses contain a specific reference to a change in the value of gas. This is not surprising, as a change in economic circumstances that has not been reflected in the price provisions will usually relate to such change.

In this section, we consider two questions that are regularly raised when a party requests a price review based on a change in the value of gas. These originate from the wording typically found in price review clauses. From this, it follows that "the value of gas" refers to: (i) the value in the market of the buyer, with or without reference to the end-user market; and (ii) to the value as it can be obtained, directly or indirectly, by a prudent and efficient gas company. This raises questions related to the applicable market level and to the obtainability of the market value of gas.

(a) Value of gas in the market of the buyer – market level

We have already discussed the geographical element that is included in price review clauses, referring to the country or the market of the buyer (or another defined relevant market) in which the change must have occurred. We have also pointed out the various questions that have arisen in this regard due to recent market developments. Apart from these more geographical questions the price review clause also raises questions related to the market level that should be used to determine the value of gas in a price review. These questions, which will be discussed in this section, revolve around the two different market levels found in most gas markets today: the end-user market and the wholesale market.

When most European long-term gas sales agreements were concluded, the gas supply chain was relatively straightforward and consisted of only a few key players. Market players bought gas from producers under long-term agreements and sold the gas onward to a variety of end-users: large industrial clients, gas-fired power plants, medium and small-sized clients. They also sold gas to households, although this was done through local distribution companies. As there was only a relatively small mark-up included in the gas prices, the distribution companies mostly served as an intermediary.

Today, part of the value of gas has become stuck at this intermediate level. As distribution companies have been able partly to buy gas from the increasingly liquid spot markets, but are – at the same time – faced with consumers reluctant to change suppliers and renegotiate better prices, they have successfully negotiated lower purchase prices compared with relatively high sales prices.

As a result, the definition of the term 'market of the buyer' has increased in importance when determining the value of gas. The value of gas can differ depending on the market level that is referred to. Buyers argue that as they mainly sell gas at the wholesale level, this is the relevant market. However, some price review clauses make explicit reference to the value of gas in the end-user market. Sellers argue that this suggests that instead of the intermediate wholesale level, the end-user level is the 'market of the buyer'. Buyers refer to the wording 'directly and indirectly' sometimes included in the price review clause to argue that the value of gas at the wholesale level should be taken into account. By contrast, sellers argue that this wording suggests that the value of gas includes the value that is obtained at the wholesale level by distribution companies when they sell gas at a higher price to households, which are considered part of the end-user market. Sellers claim that buyers obtain this value indirectly through their sales to the distribution companies (particularly if the buyer is part of a wider group of companies that includes distribution companies), the value of which should therefore also be included in the calculation of the value of gas in the buyer's market. To our knowledge, no award in which a tribunal considers the validity of either position has been published yet.

(b) Technical market value versus obtainable market value – the obtainability test

When determining the value of gas, price review clauses may refer not only to the value in the (end-user) market of the buyer, but also to the value as it can be obtained (directly or indirectly) by a prudent and efficient gas company. This final part clearly reflects that the price review clause refers to an objective value of gas, although this still leaves room for a buyer to support its value calculations with its own sales prices, as long as the buyer can be considered 'prudent and efficient'. What causes much more debate is the question of whether the value referred to in this part of the price review clause is the technical market value or the obtainable market value.

When most long-term gas sales agreements were concluded, there was no actual gas-to-gas competition. Instead, gas competed with alternative fuels. In order to ensure the long-term competitiveness of gas, its value and, thus, its price was determined by identifying the best alternative fuel for each type of end-user and pricing the gas competitively in relation to such best alternative.[21] In most cases, the best alternative was oil. Prices in long-term gas supply contracts were therefore mainly linked to oil product prices.[22] The price of gas was intended to reflect the value of gas to the end-user in comparison with the

---

21 The market value principle was introduced in 1962 by the Dutch minister of economic affairs, JW De Pous, who published a policy paper referred to as "*Nota De Pous*" (Parliamentary Papers II, 1961/1962, 6767, nr 1). See also Anthony J Melling, *Natural Gas Pricing and Its Future, Europe as the Battleground* (Carnegie Endowment, 2010) 16–17.

22 EU Energy Sector Enquiry, 10 January 2007, pp101–104 and International Energy Agency, *Natural Gas Pricing in Competitive Markets* (1998) 32.

costs that end-user incurred when using the most attractive alternative fuel. This is referred to as the 'indifference principle' (also known as the *Anlegbarkeitsprinzip*), which sets the technical market value of gas.[23] At a time without gas-to-gas competition, this was the only and accepted methodology to determine the value of gas in a price review, providing a technical market value.[24]

However, as, generally, price review clauses refer to the value as it can be obtained, this has led to parties and tribunals raising questions regarding the meaning of the term 'value of gas' and the methodology to calculate such value. Some parties have argued that the 'value of gas' refers to the market value or obtainable value of gas, also referring to the wording 'directly and indirectly', which can be included in price review clauses. They argue that this wording confirms the price review clause refers to the value of gas as it can actually be obtained. In that regard, in its final award in ICC Case 13504 the tribunal considered that:

> *in the absence of a particular provision in the [Contract] as to the meaning of "the value of Natural Gas", the Arbitral Tribunal considers that this expression should be interpreted as referring to the ordinary market value of gas, and that the market value of gas is to be determined on the basis of prices obtained or obtainable in actual transactions in the various market segments constituting the end user market of the Buyer.*
>
> *... The value obtained in the market is the payment/revenue/price which the supplier receives from the entirety of all end customers. The principle that any value determination under [the price review clause] has ultimately to be checked against and is finally determined by the prices actually obtained or obtainable in the market was acknowledged by Respondent.*[25]

The discussions on the difference between the technical market value and the obtainable market value have been fuelled by various market developments, such as the successful production of US shale gas, the increased number of continental European LNG terminals and the financial crisis. These developments have caused a simultaneous supply surge and a drop in demand, which in Europe has led to oversupply situations since around the second quarter of 2009. Combined with the increased liquidity of spot markets, gas-to-gas competition is considered to have emerged to varying extents in certain key

23 In its 1998 publication, the International Energy Agency refers to the netback value of gas, which is determined "on the basis of the weighted average value of the gas in competition with other fuels adjusted to allow for transportation and storage costs from the beach or border and any taxes on gas". See n 22 above, p32. See also Hans K Schneider and Walter Schulz, *Die Gaspreisbildung nach dem Anlegbarkeitsprinzip, Bin Beitrag zu ihrer wirtschaftspolitischen Beurteilung* (R Oldenbourg Verlag, Munich, 1977) which discusses the Anlegbarkeitsp1inzip and indexation-based pricing.

24 Final Award in Case 9812, August 1999, "Extracts from ICC Arbitral Awards: Price Setting and Price Revision in the Energy Sector" (2009) 20(2) *ICC International Court of Arbitration Bulletin* 75.

25 Final Award in Case 13504, January 2007, in "Extracts from ICC Arbitral Awards: Price Setting and Price Revision in the Energy Sector" (2009) 20(2) *ICC International Court of Arbitration Bulletin* 99.

European markets. In addition, these developments caused a spread between the prices of gas on hubs and the price of gas that is traded based on oil price indexation. These prices tended to be both linked and of similar levels over the mid to longer term, yet this has no longer been the case since the summer of 2009. The influence of this development on the actual obtainable value of gas still depends on the particular market involved in the contract. It should be noted that today, the transparent gas spot markets and decline in oil prices have caused this spread between oil and gas prices to – at times – disappear.

Following these developments, parties have argued that the wording of the price review clause allows for a shift from the valuation of gas versus alternative fuels to a valuation of gas versus gas. They argue gas to gas competition allows parties to determine the value of gas based on the value that the buyer can actually obtain in its market, which they argue is currently determined by the prices on the spot market, thus definitively moving away from the more technical market value and instead applying an obtainability test. Advocates of the obtainability test claim that the wording of the price review clause confirms that it refers to the value of gas as it can actually be obtained, which they argue is currently based on gas prices in the spot market.

3.3 Ability to economically market the gas

Usually, the price review clause contains a separate paragraph which provides that in any case the price provisions of the contract shall allow the buyer to economically market the gas. This is also referred to as the 'in any case' clause. Generally, the wording of this clause raises two questions on its interpretation:

- Does it provide a separate trigger for a price review?
- Does it provide a margin guarantee?

(a) Separate trigger?

It is arguable whether the 'in any case' clause constitutes a separate basis to trigger a price review. An argument to that extent may be made based on the phrase 'in any case' which is generally found at the beginning of the clause. However, it follows from the price review clause that a price review requires the economic circumstances to change significantly and that this is the true core of a price review. The phrase 'in any case' could simply mean that any adjustment of the price provisions reflecting these changes should leave the gas marketable.

In the limited case law that is available, reference has been made to the in any case clause and its interpretation. In an ICC award rendered in 1999, the arbitral tribunal held that:

> *Therefore, the Arbitral Tribunal concludes that a consolidation of both paragraphs has to take place. Such a consolidation entitles the buyer to invoke Article 6.10(a)(2) [the in any case clause] if an application of Article 6.10(a)(1) leads to a result where he cannot economically market the natural gas. Accordingly, the*

second paragraph therefore works as a protection of the buyer against upward adjustments of the price considering his market situation.
(...) the Arbitral Tribunal has found that an allegation of 'non-marketability' (...) cannot be used by the Claimants as an independent basis for its request for price review, but can only be invoked as a defense against the Respondents' request for price increase (...).[26]

In another ICC award rendered in 2007, the arbitral tribunal did not rule on the question whether the in any case clause can serve as a separate trigger in a price review.[27] The current case law suggests that if certain changes in the economic circumstances in principle require an upward adjustment of the contract price, the buyer may still resist such a price adjustment if this would make it impossible for it to economically market the gas. So the clause appears to serve as a form of safety net provision for the buyer, and not as a separate trigger for a price review.

It is unclear whether the in any case clause provides greater protection to the buyer. Some buyers argue that if they can identify a change that qualifies as a trigger for a price review, the resulting price adjustment not only has to reflect that change, but is also determined by the in any case clause. The adjusted price should allow the buyer to economically market the gas: if the price adjustment following from the change in economic circumstances does not allow the buyer to economically market the gas, an additional adjustment should be made. Although this interpretation may sound reasonable from a buyer's perspective, to our knowledge there is no publically available award in which a tribunal has ruled on this interpretation.

(b) Margin guarantee?

Normally, the in any case clause does not provide an explanation of the term 'economically market'. Some buyers argue that the in any case clause also provides a margin guarantee. They argue that the fact that they should have the ability to economically market the gas means that the buyer should always be able to make a margin. Clearly, the clause provides that in the end, the gas should be marketable. It is unclear whether this means that a positive margin should be obtained at all times over the duration of the contract. Based on the wording of the contract, arguments can be made both ways.

Buyers could argue that the term 'economically' refers to the buyer's ability to obtain a positive margin. Further, they could argue that whereas the intermediate price risk lies on the buyer, the seller carries the ultimate price risk

26 Final Award in Case 9812, August 1999, "Extracts from ICC Arbitral Awards: Price Setting and Price Revision in the Energy Sector" (2009) 20(2) *ICC International Court of Arbitration Bulletin* 72–73.

27 Final Award in Case 13504, January 2007, "Extracts from ICC Arbitral Awards: Price Setting and Price Revision in the Energy Sector" (2009) 20(2) *ICC International Court of Arbitration Bulletin* 97. In para 101 of its award, the tribunal does note that: "This clause is regarded by both Parties as a safeguard for the benefit of the Buyer."

reflected in the price review clause. This division of risks is addressed in the in any case clause, allowing an adjustment of the contract price so the buyer again has a positive margin at the end of the price review, ensuring that the price risk ultimately lies with the seller.

By contrast, sellers could argue that in the absence of a specific stipulation, the contract cannot be read to provide a positive margin guarantee. In addition, as it is within the buyer's control whether it obtains a margin, providing a margin guarantee would go against the first paragraph of the price review clause. In any event, arbitral tribunals and parties alike will in each case have to rely (as appropriate) on the negotiation history of the contract and previous price reviews to assess whether it can be argued that the 'in any case' clause provides a (positive) margin guarantee.

Some buyers even suggest that the in any case clause provides that they should be able to make the same margin at the start and at the end of the review period, or alternatively, that the margin at the start of the review period provides the benchmark when determining the margin at the end of the review period. Again, arguments can be made for both buyers and sellers. As a general note we point out that if certain changes in the economic circumstances changed the market conditions, it could be argued that the buyer may no longer be entitled to the same margin. However, whether a buyer would be entitled to a fixed margin largely depends on the facts of the case.

It should be noted that the 'in any case' clause typically refers to the buyer's ability to economically market the gas. In that case, there does not seem to be much room to rely on data related to competitors, or general benchmarks. Instead, the economic marketability analysis should be conducted based on the position of the specific buyer, and based on its own data. This is typically data that is accessible only to the buyer, and providing a counterparty direct access to this data will raise competition law concerns. To address these issues, it is not uncommon in price review arbitrations for the buyer to produce a report by an accountant providing data on its revenues and costs on an aggregated level. This way, the buyer can substantiate its claim and give the seller access to the relevant yet highly confidential data.

3.4 Burden of proof and the extent of the adjustment

This brings us to the burden of proof. If the formal requirements set out above have been met, this does not necessarily result in a price adjustment. The burden to prove the requested adjustment lies with the party requesting such adjustment. It will therefore also bear the evidentiary risk with respect thereto, and will have to substantiate its claim for adjustment of the price provisions by submitting expert evidence and – to the extent relevant – fact witnesses.

In case the burden of proof is not met, no adjustments can be made. Given that the 'in any case' clause seems to function as a safety net for the buyer and not as a separate basis for a price review, and given the information imparity between the buyer and the seller,[28] this would also mean that the burden of proof of the (in)ability to economically market the gas lies with the buyer.

The burden of proof also raises another question: should there be a causal link between the changes in the economic circumstances and the requested adjustment? What if only one of several changes have been proven - can that lead to a total overhaul of the price provisions? The price review clause typically provides that the price provisions are to be adjusted to reflect the changes in the economic circumstances. In our view, if only one change can be substantiated, this should lead to an adjustment reflecting only that change.

4. Some views in respect of the criteria for a price review in LNG contracts

European pipeline contracts have been at the center of most of the price review arbitrations so far. Now that European pipeline contracts have generally moved from oil-indexation to spot-indexation, the focus is increasingly shifting to LNG contracts for deliveries in Asia. LNG contracts in Asia are generally still oil-indexed and prices in Asia tend to be higher than in Europe and the United States.

Price review clauses in LNG contracts may look similar to price review clauses in contracts for pipeline gas. This was indeed the case in the well-known case of *Gas Natural v Atlantic LNG*.[29] However, price review clauses in LNG contracts tend to be more varied than in contracts for pipeline gas.

While European pipeline gas contracts generally contain a firm right to a revision of the price if certain conditions are met, this is by no means the case for all LNG contracts. LNG from the United States is generally sold under contracts that do not contain any price review clauses. Furthermore, a significant number of generally older Asian LNG contracts contain price review clauses that only stipulate that parties will meet and discuss the price, without stipulating that the price has to be revised if certain conditions occur.

There are also many LNG contracts that do contain a firm right to a revision of the price if certain conditions are met. Like pipeline contracts, these LNG contracts generally require the occurrence of a trigger in the review period. In the context of establishing whether such trigger has occurred, the value of LNG is usually key. This value may be determined on the basis of the value of gas in the country of the buyer as it is in many pipeline contracts.

However, there are other approaches as well. This may be due to the fact that

---

28 See also Mark Levy, "Gas Price Review Arbitrations: Certain Distinctive Characteristics" in *The Guide to Energy Arbitrations – Second Edition* (June 2017) available at https://globalarbitrationreview.com/chapter/1142623/gas-price-review-arbitrations-certain-distinctive-characteristics.

29 *Gas Natural Aprovisionamientos SDG, SA v Atlantic LNG Co of Trinidad and Tobago*, No. 08 Civ 1109, 2008 WL 4344525 (SDNY 16 September 2008).

LNG – being transported by vessel – has a more flexible destination than gas that is transported by pipeline. Many LNG contracts therefore seek to define a regional price as a benchmark for the price review. Certain LNG price review clauses in this respect refer to an average border price as for instance published by customs or by consultancies. Other LNG price review clauses may contain a more general reference to for instance the average price for LNG in North-East Asia. Since prices and conditions under LNG contracts are generally confidential, it often poses significant challenges to establish the level of such an average price.

5. Summary

In general, it follows from most price review clauses that a party is entitled to an adjustment of the price provisions if there are changes in the economic circumstances which:

- occurred within the review period;
- are significant compared with what is reflected in the prevailing price provisions;
- are beyond the control of the parties; and
- occurred in the country or the market of the buyer.

These requirements confirm and stress the purpose of a price review: to allow for adjustments of the price provisions so that they properly reflect the changes in the market conditions faced by the parties, thereby ensuring that the contractual balance is maintained. It is questionable whether the 'in any case' clause can be used not only as a shield or protection for the buyer if the seller requests a price increase, but also as a sword or separate trigger and margin guarantee to request a price decrease.

Changes of circumstances as a price modifier

Michael Polkinghorne
White & Case LLP

1. Introduction

In many respects, this chapter deals with matters that are foreign to the rest of this book. First, it concentrates on civil, and to an extent Arab, legal systems (whereas many gas contracts, notably LNG SPAs, are English law beasts); secondly, it considers the notion of hardship or *imprévision* (whereas we are accustomed to speaking of and dealing with gas price reopener clauses rather than more general calls for contract amendment).

The purpose of this exercise is to explore the relief open to victims of changed circumstances under the law applicable to the parties' agreement. The aim is both to demonstrate that remedies may arise independent of any gas price reopener (or in some cases, *force majeure*) clauses, and to consider how these might apply.

The starting point – whether under civil or common law principles – is that the parties are bound by their agreement.[1] However, based on the premise that the contract is not just the fruit of the parties' will but also the expression of a normative (and economic) equilibrium, courts frequently retain a residual right to intervene in cases where matters change so fundamentally as to change that agreement to something else. While parties are well familiar with the notion of *force majeure*, and much of this book is devoted to contractual price reopeners, other remedies may still lie open, and very often do. This is most notable in civil law jurisdictions.

A detailed description of the principle of hardship can be found in the UNIDROIT principles:[2]

> *SECTION 2: HARDSHIP*
>
> *ARTICLE 6.2.1*
>
> *(Contract to be observed)*
>
> *Where the performance of a contract becomes more onerous for one of the parties, that party is nevertheless bound to perform its obligations subject to the following provisions on hardship.*

---

1 While this is largely self-evident under common law systems, those unfamiliar with civil law may sometimes overlook the fact that French law, for example, echoes these sentiments.

2 Articles 6.2.1 to 6.2.3 of the UNIDROIT Principles of International Commercial Contracts (2004).

ARTICLE 6.2.2
(Definition of hardship)
There is hardship where the occurrence of events fundamentally alters the equilibrium of the contract either because the cost of a party's performance has increased or because the value of the performance a party receives has diminished, and
(a) the events occur or become known to the disadvantaged party after the conclusion of the contract;
(b) the events could not reasonably have been taken into account by the disadvantaged party at the time of the conclusion of the contract;
(c) the events are beyond the control of the disadvantaged party; and
(d) the risk of the events was not assumed by the disadvantaged party.

ARTICLE 6.2.3
(Effects of hardship)
(1) In case of hardship the disadvantaged party is entitled to request renegotiations. The request shall be made without undue delay and shall indicate the grounds on which it is based.
(2) The request for renegotiation does not in itself entitle the disadvantaged party to withhold performance.
(3) Upon failure to reach agreement within a reasonable time either party may resort to the court.
(4) If the court finds hardship it may, if reasonable,
(a) terminate the contract at a date and on terms to be fixed, or
(b) adapt the contract with a view to restoring its equilibrium.

Relevant provisions on hardship are also found in the Principles of European Contract Law[3] or in the OHADAC Principles on International Commercial Contracts.[4]

By way of comparison, examples of commonly used language in gas price reopeners include the following:

- *Either party may request a review of the method of determining the Contract Sales Price set forth in clause X if it has a good faith basis for believing that, for reasons outside its control, there has been a material change in the value of imports into Y and/or re-gasified LNG in the Y gas market which is anticipated to have lasting effect.*[5]

---

3 See Article 6:111 on "Change of Circumstances". While the UNIDROIT Principles deal with international contracts worldwide, the European Principles concern domestic and consumer contracts in the European Union. The two regimes are quite similar, with an additional rule contained in the European Principles allowing the court to "award damages for the loss suffered through a party refusing to negotiate or breaking off negotiations contrary to good faith and fair dealing".

4 See Article 6.3.1 on Hardship. The OHADAC Principles adopted in 2015 in the Caribbean, depart from the UNIDROIT model, not only because they provide for the same rules to apply both to hardship and to frustration but also in that the only remedy they provide for in case of hardship (or frustration) is for the aggrieved party to unilaterally terminate the contract.

- *If … economic circumstances in the buyer's market … have substantially changed as compared to that expected when entering into the contract for reasons beyond the parties' control … and the contract price … does not reflect the value of natural gas in the buyer's market.*[6]

Hence, while hardship allows the entire contract to be reviewed, indexation and price adaptation clauses are most frequently specific to the question of price. Nonetheless, both price review and hardship mechanisms deal with the tension – inherent in any long-term arrangement – between the opposing aims of predictability and flexibility. Parties want to know what they have to do but recognise that no one can predict the future.

Thus, both reopeners and hardship clauses look at ways of dealing with the unexpected. They provide that in the case of an unforeseen event, beyond the control of the parties, which renders performance unduly onerous (the precise terms and degrees vary), something must be done. That 'something' generally involves the parties seeking to negotiate a solution, failing which a third party (an expert, arbitral tribunal or national court) can often impose a solution by way of contract modification.

Are the underpinnings of the two measures the same? Looking at price reopeners (and this chapter concentrates on the European form), their aim may be discerned as the preservation of the contract's feasibility, which from the outset involves the buyer's ability to market the gas in its end market(s). The clauses involve the recognition that the end game – the ability to market the gas – is one of the *raisons d'être* of the entire arrangement.[7]

Hardship provisions, on the other hand – at least insofar as they are found in national legal systems – are not nearly as specific. They are geared to apply to

---

5 P Hodges, "LNG – a minefield for disputes?" in P Griffin (ed), *Liquefied Natural Gas: The Law and Business of LNG* (Globe Law and Business, 2006) 115.

6 S Farmer, "LNG Sale and Purchase Agreement" in P Griffin (ed), *Liquefied Natural Gas: The Law and Business of LNG* (Globe Law and Business, 2006) 49. See also an example from an Energy Charter Secretariat paper (Energy Charter Treaty Paper, "Putting a Price on Energy" (18 October 2007) at 161 (available from the Energy Charter Treaty website):

(a) If the circumstances beyond the control of the Parties chance [sic] significantly compared to the underlying assumptions in the prevailing price provisions, each Party is entitled to an adjustment of the price provisions reflecting such changes. The price provisions shall in any case allow the gas to be economically marketed based on sound marketing operation.

(b) Either Party shall be entitled to request a review of the price provisions for the first time with effect of dd/mm/yyyy and thereafter every three years.

(c) Each Party shall provide the necessary information to substantiate its claim.

(d) Following a request for a price review the Parties shall meet to examine whether an adjustment of the price provisions is justified. Failing an agreement within 120 days either Party may refer the matter to arbitration in line with the provisions on arbitration of the Contract.

(e) As long as no agreement has been reached or no arbitration award has been rendered all rights and obligations under the agreement – including the price provisions – shall remain applicable unchanged. Unless otherwise agreed or decided by the arbitral award, differences to the newly established price shall be retroactively compensated inclusive of interest on the difference calculated at a rate reflecting the conditions on the international financing market.

7 Another factor being that the seller receives sufficient revenues to ensure the long-term viability of the deal. While one often speaks of buyers taking volume risk and suppliers price risk, the issues are often more nuanced than this simplified description might suggest.

an undisclosed variety of events, and look more broadly to their overall effect on the parties rather than the commercial viability of the deal struck (although this aim can be the subject of a hardship provision).

Also, as explained in ICC Case 3344, the main difference between a price adaptation clause and a hardship clause lies in the automaticity of the adaptation of the price: the parameters of adaptation set out in the contract need generally to be objective and lead to a single solution.[8]

Questions of the parties' positions and commercial viability are closely related, particularly when price reopeners speak of adjustments that are "fair and equitable" or refer to contractual "equilibrium"; but the fact remains that they are different animals. The question for this chapter is thus whether – in the face of a well-drafted price reopener – a role remains for hardship principles to operate, and if so, when and how.

And what if there is no price reopener clause at all? What options are open to a party to a long-term contract that is no longer viable?

2. Hardship under national laws

2.1 Common law notions?

It seems fairly settled that English law will not admit a (non-contractual)[9] power to amend a contract on the basis of changed circumstances.[10] Perhaps the closest one might come to this is the doctrine of frustration, which – while arguably broader than the notion of *force majeure*[11] – results in termination rather than modification of the underlying obligation.[12]

One area which remains the subject of some debate is the application of good faith in long-term contracts; put another way, can an obligation to negotiate in good faith be enforced *per se*, and/or could a tribunal take the analysis even further to impose what it would have expected as a negotiated solution had the party indeed negotiated in good faith?

Not surprisingly, this is a big 'ask', but it is not a sterile debate. For example, the 2013 *Yam Seng* case[13] saw Mr Justice Leggatt (as he then was) revisit the

---

8 Award rendered in 1981, see S Jarvin and Y Derains (eds), *Collection of ICC Arbitral Awards 1974–1985* (Kluwer Law & Taxation/ICC Publishing, 1990) 440 cited in G Block, "Arbitration and Price Renegotiation in Energy Contracts" (2017) 2 *ICC Dispute Resolution Bulletin* 52.

9 For an example of an English court's treatment of a contractual hardship clause, see *Superior Overseas Development Corporation v British Gas Corporation* [1982] 1 Lloyd's Rep 262. (A French example can be seen in CA Paris, December 28 1976, *EDF v Shell France*, JCP G 1978, II, 18810, obs J Robert.)

10 See, eg, *Total Gas Marketing v Arco British Ltd* [1998] 2 Lloyd's Rep 209 (House of Lords).

11 *Davis Contractors v Fareham Urban DC* [1956] UKHL 3, [1956] AC 696, [1956] 2 ALL ER 145 cited at AH Puelinckx, "Frustration, Hardship, Force Majeure, Imprévision, Wegfall der Geschäftsgrundlage, Unmöglichkeit, Changed Circumstances" (1986) 3(2) *J Int'l Arb* 47 *et seq*, at p50: "frustration occurs whenever the law recognises that without default of either party a contractual obligation has become incapable of being performed because the circumstances in which performance is called for would render it a thing radically different from which was undertaken by the contract... it was not this that I promised to do".

12 *ibid.*

established position of good faith under English law – he asserted that the general view in English contract law, that there is no overriding principle of good faith, is "swimming against the tide", and even though he doubted that English law was ready to recognise good faith as a duty implied by law into all commercial contracts, he found that there was "nothing novel" in implying such a duty into an ordinary commercial contract, based on the presumed intention of the parties. Mr Justice Leggatt went on to emphasise the importance of recognising such a duty in long-term agreements which require a high degree of communication, cooperation and predictable performance based on mutual trust, confidence and loyalty. These bases need not be set out expressly in the contract but – he states – are implicit in the parties' understanding. In fact, in a lecture given recently, Lord Justice Leggatt (as he now is) specifically employed the hypothetical case of a change of circumstances in a contract for the supply of Liquefied Natural Gas: he did so to illustrate ways in which "legal and practical content could be given to a promise to negotiate in good faith".[14] The *Yam Seng* reasoning was recently cited[15] or explicitly approved[16] by other common law courts. Interestingly, in 2015, an ICC tribunal found that the termination of a gas sale and purchase agreement was wrongful because of a breach of an implied duty of good faith, even though the parties had argued that there was no such obligation in the contract or under English law governing the agreement.[17] Although this is an investment case, it is interesting to note that the tribunal found the fact that the deal had been made between sovereign states, entailing substantial quantities of gas and significant investment by all parties, to be "particularly compelling".

This developing case law seems to resist implying a duty of good faith in the contractual negotiation phase but seasoned commentators have suggested that the common law does appear ready to move (or start moving?) towards a default rule of performance of certain contracts in good faith.[18]

The United States has a doctrine of commercial impracticality, pursuant to which several commentators have posited that the court may not only arrive at a fair distribution of losses, but also modify the parties' agreement going forward.[19] That said, the courts appear to have been parsimonious in the exercise of such powers.[20]

---

13 *Yam Seng Pte Ltd v International Trade Corporation Ltd* [2013] EWHC 111. The Court concluded that the distribution agreement in question contained an implied and enforceable obligation on the parties to act in good faith, and that International Trade Corporation had knowingly concealed from Yam Seng its true pricing arrangements and a crucial distribution channel in Singapore, thus causing a repudiatory breach justifying Yam Seng's termination of the distribution agreement.

14 G Leggatt, "Negotiation in Good Faith: Adapting to Changing Circumstances in Contracts and English Contract Law", Jill Poole Memorial Lecture, Aston University, 19 October 2018, paras10–59.

15 *The One Suites Pte Ltd v Pacific Motor Credit Ltd* [2015] SGCA 21.

16 *Globe Motors Inc v TRW Lucasvarity Electric Steering Ltd* [2016] EWCA Civ 396; *National Private Transport Services Company Limited v Creditrade LLP* [2016] EWHC 2144 (Comm).

17 D Charlotin, "ICC Tribunal finds that international agreement includes an implicit good faith obligation", IA Reporter, 7 December 2018.

18 P Griffin, "Principles of price reviews and hardship clause in long-term gas contracts" in *Liquefied Natural Gas: The Law and Business of LNG* (Globe Law and Business, 2006) 122.

One might also pause to discuss the notion of *force majeure* (which does not exist *per se* under English law – hence the large quantities of ink often spilled in drafting these clauses).

Like the above notions, *force majeure* looks to deal with the unpredictable, but it has its limitations. In the usual case, the occurrence of an event of *force majeure* results in either suspension or termination of the agreement in question (although examples exist whereby the parties undertake to negotiate an accommodation). In addition, the requirement under most *force majeure* provisions is that the supervening event render contractual performance not merely onerous, but impossible. In this regard, there is a fairly consistent stream of authority that problems in end-market conditions are not of a type to render a purchase obligation impossible to perform – the usual test for an event of force majeure.

However, as *force majeure* is a contractual phenomenon under the common law, skilful drafting of a *force majeure* clause may lead to a different result. Two examples may illustrate this. One involves the use of the word 'hinder'. It is generally accepted under English law that where a clause refers to performance being hindered, this is a lesser standard than impossibility – as was stated nearly one century ago: "to place a merchant in the position of being unable to deliver unless he dislocates his business and breaks his contracts in order to fulfil one surely hinders delivery."[21] So the use of such language may permit a court to intervene in circumstances falling short of impossibility to perform (although even here a price rise may not suffice).[22]

The other is an example of pure wordsmithing. While end-user market problems would not usually fall into the category of *force majeure*, there are examples of cases where *force majeure* has been established purely due to the fact that the clauses defining the same included language that temporary loss of markets would be covered by the relevant clause.[23] One thus needs to take care in drafting and reviewing clauses of this nature.

---

19 For a general discussion, see Fred Fucci, "Hardship and Changed Circumstances as Excuse for Non-Performance of Contracts – Practical Considerations in International Infrastructure Investment and Finance", provisional TDM, January 2007.

20 See *Aluminium Co of America v Essex Group Co* 499 F Supp 53 (WD Pa 1980) cited by Fucci (id) at p2, not followed in later cases; see White & Peters, "A Footnote for Jack Dawson" 2002 University of Michigan Law School, available at https://repository.law.umich.edu/articles/848. See also *Florida Power & Light Co v Westinghouse Electric Corp* (517 F Supp 440 (ED Va 1981)), in which the court noted: "Generally, however, the fact that performance has become economically burdensome or unattractive is not sufficient for performance to be excused… We will not allow a party to a contract to escape a bad bargain because it is burdensome."

21 *Tennants (Lancashire) Ltd v CS Wilson & Co Ltd* [1917] AC 495 at 510. Cited in *Chitty on Contracts*, 31st edn, para 14.146.

22 *ibid.*

23 See *Kodiak 1981 Drilling Partnership v Delhi Gas Pipeline Corp*, No 83-CI-10260 (Bexor Co Tex 1984), cited in WJ Legg, *Natural Gas Contract Litigation in Oklahoma*, 11 Okla City UL Rev 63 (1986) at 78.

2.2 Civil law

In the civil law context, however, the notion of hardship is found in a surprisingly large number of systems and may spring from several bases. The vast majority of examples cited are set out in the annex to this chapter.

(a) Arab law countries

In many Arab countries, the theory of revision for hardship finds its true roots in Islamic law. As a religious system, Sharia is concerned above all with questions of individual conscience, good conscience and good faith. Thus, to hold the debtor to performance in circumstances in which it had never envisaged performance would – the argument runs – be unconscionable and inequitable.[24]

According to a general principal of Islamic law, "hardship shall bring alleviation". This general principle of Islamic law emanates from the *Quranic* principle that "Allah intends for your ease and does not intend for your hardship",[25] and from a *hadith*[26] reporting that the Prophet Muhammad said that "you have been sent forth to make things easy, not to impose difficulties".[27]

Hanafi[28] jurists have elaborated a 'doctrine of excuses' (*i'dhar*) under which the aggrieved party is excused from performing its contractual obligations – and the contract may be terminated or modified – if performance becomes prejudicial subsequent to the occurrence of an unforeseen circumstance after conclusion of the contract.[29] *Hanbali*[30] and *Maliki*[31] jurists adopted a 'doctrine of calamity/blight' (*ja'iha*), which is similar to the *Hanafi* doctrine of excuses.[32]

Egypt, followed by other Arab countries, adopted the theory of hardship when incorporating much of the French Civil Code in the 1940s. Article 147(2) of the Egyptian Civil Code provides as follows:

- *The contract makes the law of the parties. It can be revoked or altered only by mutual consent of the parties or for reasons provided for by the law.*
- *When, however, as a result of exceptional and unpredictable events of a general character, the performance of the contractual obligation, without becoming*

---

24 D Rivkin, "Lex Mercatoria and Force Majeure" in *Transnational Rules in International Commercial Arbitration* (ICC Publ No 480/4, 1993) 196.

25 Sûrah al-Baqarah, 185. The following quotations involve informal translation.

26 A report of the sayings or actions of the prophet or his companions.

27 Sahîh al-Bukhârî, Vol 1, Book 4, No 219.

28 The Hanafi school is one of the four main Sunni schools of law. It was founded by scholar Abu Hanifa an-Nu'man and is followed by most Sunnis of Iraq, Syria, Lebanon, Bangladesh, Mauritius, Pakistan, India, Afghanistan, Central Asia, Southern Russia, the Caucasus, the Balkans, Turkey and parts of Egypt.

29 A Sanhoury, *Al Waseet fi Sharh al Qanun al Madani*, Vol 1 (2008) 525–26, para 416; J Whelan, *UAE Civil Code and Ministry of Justice Commentary*, p156, para 2-0273.

30 The Hanbali school is one of the four main Sunni schools of law. It was founded by scholar Ahmad ibn Hanbal and is considered very strict and conservative, especially regarding questions of dogma and cult. The Hanbali school is followed by most Sunnis of Saudi Arabia.

31 The Maliki school is one of the four main Sunni schools of law. It was founded by scholar Malik ibn Anas and is followed by most Sunnis of North Africa, West Africa, the United Arab Emirates and Kuwait.

32 J Whelan, *UAE Civil Code and Ministry of Justice Commentary*, p156, para 2-0273.

> *impossible, becomes excessively onerous in such way as to threaten the debtor with exorbitant loss, the judge may, according to the circumstances, and after taking into consideration the interests of both parties, reduce to reasonable limits, the obligation that has become excessive. Any agreement to the contrary is void.*[33]

In North Africa, this is echoed in Article 107(3) of the Algerian Civil Code and Article 147 of the Libyan Civil Code, among others, however, it appears that Morocco and Tunisia have not adopted this provision.

In the Middle East, one can look to Bahrain (Article 130 of Legislative Decree 19/2001), Iraq (Article 146(2) of the Civil Code), Jordan (Article 205 of the Civil Code), Kuwait (Article 198 of the Civil Code), Qatar (Article 171(2) of the Civil Code), Syria (Article 148 of the Civil Code), the United Arab Emirates (Article 249 of the Civil Code) and Yemen (Article 211 of Civil Law 14/2002).[34] Lebanon has limited the application of this principle to administrative (or public law) contracts.

(b) Europe

In Europe, the theory has ebbed and flowed over the centuries. Founded on Roman law, the principle of *rebus sic stantibus* (ie, the contract remains binding provided that matters remain as they are) enjoyed some popularity in the 18th century, but fell out of favour in the 19th century. It returned to the fore in the early 20th century, as reflected in the French *Gaz de Bordeaux* case.[35]

In this case, a concessionaire for the provision of gas lighting in Bordeaux found that the price of coal had tripled, well outside any contractual range. The French Conseil d'Etat (the highest administrative court in France) ultimately ruled that the city of Bordeaux should compensate the concessionaire on the basis of this unforeseen event (*'imprévision'*).[36] As a result, French administrative law has traditionally admitted the notion of hardship, although until very recently, the same principle did not exist in private law. It was only in 2016 that the more generally applicable theory of *imprévision* was introduced in the French Civil Code by a reform of contract law; but more on that later.[37]

---

33 Perrot Fanner and Sims Marshall, *Translation of the Egyptian Civil Code* (New World Publisher) 29.
34 See Annex 1, section 5 below.
35 Conseil d'Etat, Compagnie générale d'éclairage de Bordeaux, Rec 125, concl Chardenet, 30 March 1916, quoted in M Long, P Weil et al, *Les Grands Arrêts de la Jurisprudence Administrative* (2003) 188–89. Translation from the award in ICSID Case ARB/01/8, *CMS Gas Transmission Company v The Argentine Republic* (12 May 2005); all cited in Fucci (see fn 19), p18.
36 *ibid*: "Just as the Company cannot argue that it should not be required to bear any increase in the price... it would be totally excessive if it is admitted that such increases are to be considered a normal business risk; on the contrary, it is necessary to find a solution that puts an end to temporary difficulties, taking into account both the general interest ... and the special conditions that do not allow the contract to operate normally ... ; to this end it is necessary to decide, on the one hand, that the Company is required to provide the concession service and, on the other hand, that during this period it must bear only that part of the adverse consequences that a reasonable interpretation of the contract allows ...".
37 See section 4.3(b) below.

Other countries have long since recognised the principle as one of general application. For example, Article 6:258 (complemented by Article 6:260) of the Dutch Civil Code and Section 313 of the German Civil Code deal expressly with the effects of unforeseen circumstances in the commercial sphere, giving the court the power to modify or terminate the contract. Austrian law has a similar approach, albeit on a different basis (as in Germany, based on the so-called *Wegfall der Geschäftsgrundlage* (ie, the disappearance of the basis of the transaction)). Similar provisions are contained in the Portuguese and Greek Civil Codes (Articles 437 and 388 respectively). Along the same lines, Italian case law has interpreted Article 1467 of the Italian Civil Code as giving the courts the power to adapt the contract to the new circumstances.

As is outlined below, Belgium, and France until very recently, and Quebec till this date, have displayed reticence in giving hardship a more limited role.[38]

(c) *Nordic countries*

One can also find hardship provisions by way of Article 36 of the Contracts Acts of Sweden, Finland and Denmark.[39]

(d) *South America*

In South America, the Civil Codes of Argentina, Peru and Paraguay expressly admit change of circumstances as a ground for relief for the affected party.[40] The exception to this rule appears to be Chile, where the hardship doctrine has not been embraced by the courts outside the area of administrative contracts.

(e) *Eastern Europe*

Provisions on hardship can also be found in Eastern European jurisdictions.[41] For example, Article 451 of the Civil Code of the Russian Federation provides that if an "essential change of circumstances" *which could not have been foreseen at the time the contract was entered into and which could not be overcome by the parties using a reasonable degree of care, disrupts the equilibrium of the contract and deprives a contractual party who had not assumed such risk, of the benefit of the contract, such party may request the court to adapt the contract. Such an amendment can be ordered by the court only in "extraordinary cases" where termination of the contract would be against public interest or would entail extraordinary losses for the parties (which exceed the cost of executing the contract as amended by the court).*[42]

The Estonian Law of Obligations, in force since 2002, includes a provision on "Alteration of balance of contractual obligations".[43] *It requires a change of*

---

38 See section 4.3(a), (b) below.
39 See Annex 1, section 1 below.
40 See Annex 1, section 3 below.
41 See Annex 1, section 7 below.
42 Article 451(4) of the Civil Code of the Russian Federation.
43 See §97 of the Estonia Law of obligations.

circumstances where a material change occurs "in the balance of the obligations of the parties", due to an increase of the costs to perform or a decrease in the value of the counterpart to be received from the other party. This is generally seen, unsurprisingly, as application of the imprévision approach.[44]

Lithuania (Article 6.204 of the Civil Code) provides another example.

(f) ***Far East***

In Japan – even though there is no statutory law governing changed contractual circumstances – a judge can adapt the contract if he finds unforeseen, substantially changed circumstances which are not the fault of the parties and which mean that continued performance would violate the principles of good faith and equity.

In his book *The Principle of Changed Circumstances in Civil Law* (1926), Katsumoto Masaaki gave the following definition of the rule:

> *The rule states that when the circumstances releasing a legal act, creating a legal relation, change in an unforeseeable way alter the accomplishment of the act and before expiry of its consequences, due to the influence of facts not attributable to the parties, with the result that the creation of legal consequences in the original sense, or their continuation, is to be considered as unreasonable in the light of the rule of equity and good faith, then such legal consequences are to be adapted in equity and good faith.*[45]

In China, although contract law contains no specific provision on change of circumstances, the principle has been recognised by the judiciary, as evidenced by the "Interpretation II of the Supreme People's Court on Several Issues concerning the Application of the Contract Law of the People's Republic of China" issued by the Supreme People's Court of the People's Republic of China in 2009. Article 26 of Interpretation II provides that where material changes in objective circumstances, which are unforeseeable at the time of the conclusion of the contract, and which are not caused by *force majeure* and are not commercial risks, occur after the contract has been concluded, then if continuation of performance of the contract will be obviously unfair for one party or if the contractual purposes cannot be realised, and a party files a request with the people's court for the modification or rescission of the contract, the people's court shall determine whether to modify or rescind the contract under the principle of fairness and in light of the circumstances of the case. Article 26 requires courts to apply the principle with great prudence. After a case-by-case analysis, if a court deems the application of the principle to be inevitable, the trial court must first report to the high court at the provincial level for

---

44 See Annex 1, section 7 below.

45 K Igarashi, *Keiyaku to jijohenko*, 3rd edn (Tokyo, Yuhikiaku, 1983) 147–48 cited at AH Puelinckx (see fn 11) 62–63.

examination and approval. When necessary, the high court may report the case to the Supreme Court for examination and approval.[46]

3. Issues of hardship

While one can speak of general principles of hardship, it should be readily apparent that the manner in which these principles are applied will vary not only between different jurisdictions, but even within those jurisdictions themselves. A discussion of the most common issues of relevance to gas price matters may assist.

3.1 Nature of triggering event

(a) Need the underlying event be general or specific?

Article 147(2) of the Egyptian Civil Code, along with many other national codes in North Africa, provides that the event must be not only exceptional, but also 'general' in nature. This means that it should not be specific to the debtor but should affect a category of persons. Thus, the debtor's individual bankruptcy or a strike of its employees will not ordinarily be considered as a general exceptional event on which the debtor may base an application for hardship.[47]

However, the converse appears to be the case under Hungarian civil law. The Hungarian courts have seemed to accept that Section 241 of the old Civil Code (which reflected the hardship principle and which is now replaced by Section 6:192 of Act V of 2013 of the Hungarian Civil Code) cannot be applied to adapt contracts to deal with fundamental changes in the economic and social climate affecting a wide range or the majority of economic actors. The courts have explained that a change in market demand in itself could not serve as ground to amend a contract, since both parties were affected by the change in that case.[48]

(b) Are price changes foreseeable?

The notion of what is foreseeable will vary from case to case, and the issue of price variations may evoke different views from one jurisdiction to the next. The Federal Supreme Court of Iraq, for example, has ruled that a debtor could not rely on an increase in the general pricing level of raw materials and wages to request the amendment of its contract, since "the increase of prices is a predictable event that cannot be considered as an exceptional general

46 S Ning, T Huang and Y Yang, "New Provision Could Change Contract Law in China but Judges Remain Conservative" in *China Law Insight*, 17 November 2010.

47 A Sanhoury (see fn 29) 532, para 420.

48 Nochta, "A gazdasági válság mint szerz dési kockázat" (The economic crisis as contractual risk) in Tamás Nótári, (ed), *Ünnepi tanulmányok Sárközy Tamás 70. születésnapjára* (Lectum, Szeged, 2010) 211; BD 200.277 cited in T Papp, "Connection between the economic crisis and contractual circumstances in Hungary and in the European Union" (2012) 2 *Juridical tribune* 114–120, p117.

unforeseeable event".[49] (One wonders whether the court might nevertheless have accepted the idea that, even here, an increase of unprecedented proportions might have qualified for correction.)

Fucci[50] cites the following as examples of events which have been ruled to have been foreseeable and hence did not give rise to a claim for hardship:

- dramatic changes in market prices for products;[51]
- unfavourable general economic circumstances in a country;
- currency fluctuations, even severe;
- failure of a central bank to grant authorisation to pay in foreign currency when foreign exchange control regulations were in place at the time of contracting; and
- armed hostilities between countries with a history of antagonism.

While the above is undoubtedly true, the inquiry may nevertheless focus not merely on whether an event itself was foreseeable, but also on whether its gravity was foreseeable. As Fucci himself points out later in his article, the following events appear to have constituted events of hardship (and were hence, by definition, unforeseeable):

- a tenfold increase in price due to a post-contracting change in government regulation[52] and the imposition of new safety regulations after conclusion of a contract to build a plant requiring installation of additional equipment and "making [the contractor's] performance substantially more onerous";
- a sudden devaluation of a country's currency (the currency of payment) in the order of 80%;
- the disappearance of a market for a contracted product (as opposed to a contraction in the market for a product).[53]

---

49 Federal Supreme Court of Iraq, 7 December 2009, Decision 1289.

50 *ibid.*

51 Fucci (see fn 19) cites a case in support: "In the ICC Award in Matter No. 2216 (1974), the case arose because a Norwegian company failed to perform a petroleum purchase agreement because the price of petroleum had gone down by more than half between the time of contract and scheduled delivery. The Tribunal did not permit an adjustment in the contract price on the basis of a hardship argument. This was not found to constitute a 'bouleversement' of economic circumstances, but a simple fluctuation in market prices. In a commentary to this case by Yves Derains, then Secretary General of the ICC Court of Arbitration, he pointed out that to accept this type of fluctuation as a ground to exonerate the non-performance of a contract 'would put in danger the security of transactions'." (Yves Derains, *Chronique des sentences arbitrales* (Clunet 1975) p921.)

52 Fucci (see fn 19) states: "In that situation a party generating waste entered into a four-year contract with a waste disposal company in a certain country. The price is fixed per ton over the term of the contract. Two years into the performance, the Government of the country prescribes prices for storing waste that are ten times higher than before."

53 Fucci (see fn 19) states: "In that case, a dealer in electronic goods in the former East Germany contracted in September 1989 to buy inventory from a supplier in another former socialist bloc country. The goods were to be delivered in December 1990. In November 1990, the buyer in East Germany informed the seller that the goods were no longer of any use to it on the grounds that after unification with West Germany, there was no longer any market for the electronic products from the socialist country supplier."

It seems clear, to the author at least, that each case must turn on its own facts.

(c) *Nature of impact of that event*

Professor Sanhoury suggests that while there is no objective criterion in Egypt that allows one to know when the performance of an obligation has become "excessively onerous",[54] there does appear to be an element of objectivity when considering whether a given event could constitute hardship (rather than a specific enquiry into the effect of the relevance of an event on a given claimant). According to Sanhoury, it does not appear to matter that the debtor has substantial resources which would allow it to bear the loss with no great pain. If the debtor is threatened with a loss that is several times as large as an ordinary loss resulting from such a contract, the requirement of an excessive burden is met, he states – even if such a burden is not great in relation to the debtor's overall wealth.[55]

However, this seems to be at odds with certain authorities discussed in the European or US context. By way of example, in a case governed by Swiss law, the tribunal expressed the view that if the contract in dispute constituted only a small part of the seller's overall revenues, it would be difficult for hardship to apply (which, in the writer's view, suggests a subjective test).[56] Fucci goes on to state:

> *Therefore, it is possible to extract from this award another principle that a contract affected by a change in economic circumstances must represent a significant portion of a company's revenues for performance to be considered "unbearable". Other arbitral jurisprudence supports the theory that the standard for judging what is bearable is subjective. What is unbearable or "excessively onerous" for a relatively small company for which a particular contract represents a significant part of its revenues may be merely onerous and thus bearable for a large, well-funded company which does not risk serious financial difficulty as a result of continued performance of the contract.*[57]

As the citation from Sanhoury shows, this point appears far from settled. Even under Swiss law, other commentators have posited that the test is in fact objective, and that the fact that a particular seller is ruined is not in itself sufficient to make a successful hardship claim.[58]

---

54 A Sanhoury (see fn 29), para 420.

55 *ibid.*

56 ICC in Case 2508 (1976), reported in *Journal du Droit International No 4* (October/November/December 1977), 939.

57 Fucci (see fn 19), p26."

58 H Van Houtte, "Changed Circumstances and Pacta Sunt Servanda", p113 4(5) TDM, September 2007 – thus advancing the position that under Swiss law, one must look at whether the event in question would seriously affect anyone in the claimant's position.

(d) *Extent of changes required to trigger hardship*

One example of where hardship was allowed to operate comes from an Algerian law arbitration in the 1990s. In this case, brought under the United Nations Commission on International Trade Law Rules, the tribunal had to rule under a dollar-denominated contract entered into by an Italian construction company with the government of Kuwait for the construction of a new embassy in Algeria. A majority of the three arbitrators accepted that a depreciation of about 35% in the value of the US dollar with respect to the Italian lira constituted a changed circumstance justifying compensation to the contractor (whose costs were predominantly in Italian lire). However, the award contained a strong dissent by one of the arbitrators, who felt that the depreciation was foreseeable. He also maintained that the majority's reading of the scope of the depreciation did not meet the rigorous demands of the Algerian Civil Code provision.[59]

As has been seen above, different cases have given rise to a range of apparently contradictory outcomes. This could be the result of the unique nature of each case when all matters are considered, or could reflect differing judicial attitudes as to when relief should be granted.

Unsurprisingly, there are variations within Europe itself. The Dutch provisions, in speaking of "reasonableness or fairness", appear fairly liberal in respect of the trigger for modification; whereas other systems tend to reflect more closely the UNIDROIT notion that the change must be fundamental before one can speak of contract change.[60]

Other countries have sought to close this debate. Sudan, for example, even puts a figure on the requisite degree of pain, requiring that "the loss exceeds one third of the obligation".[61]

(e) *Changes invoked in prior proceedings*

Can a party invoke a change of circumstances that had already been invoked in a prior arbitration? In a 2015 ICC case, the tribunal rejected the respondent's position that a change that had been considered before could not be considered again. Hence, the tribunal had to determine whether the situation had changed in a significant manner because of new circumstances since the decision of the previous arbitral tribunal, and not since – as had been argued – the conclusion of the contract.[62] It would thus be open to a claimant to prove that this change

---

59 Description of In the Matter of Arbitration Proceedings between Icori Estero SpA and Kuwait Foreign Trading Contracting & Investment Co, published in (1994) 9 *Int'l Arbitration Report*, A-1; found in Fucci (see fn 19), p21.

60 That said, the preparatory works to the Dutch provisions suggest that this remedy should nevertheless be applied sparingly. See Memorie van Toelichting pp 228, 772–73; C Van Zeben, J du Ron and M Olthof, *Parlementaire geschiedenis van het nieuwe burgerlijke wetboek*, Boek 6, (Algemeen gedeelte van het verbintenissenrecht, Kluwer, Deventer, 1981) 967 at 972; cited by Houtte (see fn 58) 114.

61 See Annex 1, section 6 below.

62 See ICC Case 19374 cited in G Block, "Arbitration and Price Renegotiation in Energy Contracts" (2017) 2 *ICC Dispute Resolution Bulletin* 58.

had evolved since the previous proceedings and that the way in which it had evolved was unforeseen by the previous tribunal or court.

(f) *Relevance of assumption of risk*

In European doctrine and jurisprudence, one frequently sees discussion of the importance of a party's assumption of risk (ie, the choice of US dollars as the contract currency indicating assumption of risk of currency movements).[63] Indeed, this is encapsulated in Article 6.2.2 of the UNIDROIT principles,[64] and the Supreme Court of Canada only recently stated that based on the wording of Article 6.2.2, "unforeseeability cannot be relied on where it is clear that the party who was disadvantaged by the change in circumstances had accepted the risk that such changes would occur".[65] This may be interesting to consider in the light of the Arabic codified hardship provisions, which purport to override contractual provisions negating hardship (one can, for example, see a debate arising as to whether allocation of risk constitutes "agreement to the contrary" under the codes).[66]

(g) *Nature of request*

Under certain European and South American systems, the mechanism involves a requirement that the disadvantaged party first seek termination of the contract, which can then be avoided only if the other party proposes or agrees to modify the same instead.[67] While one might assume that, in reality, the distinctions between requests for dissolution versus modification could become blurred, one can see cases where a party seeking a price reduction would nevertheless wish to keep the contract alive and would accordingly be loath even to suggest termination.

(h) *Is consensus needed before contract can be modified?*

Under Japanese law, while some authors assert that a judge can adapt the contract against the will of the parties, the majority opinion appears to be that if, upon a request for adaptation, the other party refuses, the contract will most likely be dissolved.[68] This could well be at odds with the situation in many other

---

63 In a case arising from the Asian financial crisis, an arbitral tribunal – dealing with the catastrophic depreciation of the Indonesian rupiyah and an economic crisis of profound proportions – held, among other things, that "by pricing in US dollars rather than in Indonesian Rupiyah, the Parties unambiguously allocated the risk of a depreciation of the local currency to PLN": "When stipulations like these appear in a long-term agreement like the ESC, with respect to which it is obvious that the surrounding circumstances may change dramatically during the life of the contract, one can only conclude that the allocation of risk is intentional, indeed emphatic." See AJ van der Berg (ed), *Yearbook Commercial Arbitration*, Vol XXV (2000) 11–432, paras 194 and 195.

64 See section 1 above.

65 *Churchill Falls (Labrador) Corp v Hydro-Québec* (2018) SCC 46, para 89.

66 See Article 147(2) of the Egyptian Civil Code at Annex 1, section 5 below.

67 See Article 478 of the Brazilian Civil Code at Annex 1, section 3 below.

68 Japanese authority also suggests that the mere presence of a hardship clause does not exclude the operation of general principles. See Arbitration Court of the Japan Shipping Exchange, award dated 20 September 1975, Yb Comm Arb, 1983, 153 at 154.

countries, in which the court's ability to modify an agreement appears not to be limited by the parties' opinions.[69]

4. Is there an inherent duty to renegotiate?

A broader question arises as to whether – for example, on the basis of the civil law concept of good faith – there exists an inherent duty to renegotiate long-term agreements. While this notion may come as a surprise to contract drafters, it has provoked a good deal of discussion among arbitrators, academics and counsel.

4.1 Judicial hesitation

The first point – as explained by Klaus Peter Berger – is that arbitrators remain loath to imply such a duty where the parties have addressed the issue of unforeseen changes (ie, by virtue of a *force majeure* clause), but have not provided for such an exercise for lesser evils.[70] He cites an *ad hoc* award – from an investor-state case – which sums up this school of thought:

> *It is not for the Arbitral Tribunal to question the motives or judgment of the Parties, but to assess their rights and obligations in light of their legally significant acts or omissions. That is all; that is enough. To go beyond this role would be to betray the legitimate expectations reflected in the Parties' agreement to arbitrate, and indeed to impair the international usefulness of the arbitral mechanism...*
>
> *The arbitrators cannot usurp the role of government officials or business leaders. They have no political authority, and no right to presume to impose their personal view of what might be an appropriate negotiated solution. Whatever the purity of their intent, arbitrators who acted in such a fashion would be derelict in their duties, and would create more mischief than good. The focus of the Arbitral Tribunal's inquiry has been to ascertain the rights and obligations of the parties to the particular contractual arrangements from which its authority is derived.*[71]

Hans van Houtte[72] cites another arbitrator's views:

> *As a general rule, one should be particularly reluctant to accept it when there is no gap or lacuna in the contract and when the intent of the parties has been clearly expressed... Caution is especially called for, moreover, in international transactions where it is generally much less likely that the parties have been unaware of the risk of a remote contingency or unable to formulate it precisely.*[73]

These observations – dealing with the question of an inherent right to

69 Article 147(2) of the Egyptian Civil Code, cited in full at Annex 1, section 5 below, states: "When, however, as a result of exceptional and unpredictable events of a general character, the performance of the contractual obligation, without becoming impossible, becomes excessively onerous in such way as to threaten the debtor with exorbitant loss, the judge may, according to the circumstances, and after taking into consideration the interests of both parties, reduce to reasonable limits, the obligation that has become excessive. Any agreement to the contrary is void."

70 KP Berger, "Renegotiation and Adaptation of International Investment Contracts: The Role of Contract Drafters and Arbitrators", 36 *V and JTL* 1347.

71 *ibid* at 1353, citing UNCITRAL award of May 4 1999, 25 YB Comm Arb 13, 61 (2000).

72 H van Houtte, (see fn 58).

73 ICC Award 1512 (1971), Yb Comm Arb, 1976, 128 at 129.

renegotiate – are probably of less impact where the principle is nevertheless espoused in the underlying applicable law.

4.2 Discussion of duty to renegotiate

Another school of thought asserts the existence of such an obligation on the basis of the civil law notion of good faith. Other commentators of a more Latin bent argue that the obligation is the offspring of the *rebus sic stantibus* principle,[74] which should temper the overriding notion of pacta sunt servanda (ie, agreements must be kept). Berger cites, among other things, the UN Draft Code of Conduct for Transnational Corporations, which states:

> *In the absence of contractual clauses providing for review or renegotiation, transnational corporations should respond positively to requests for review or renegotiation of contracts concluded with Governments or governmental agencies in circumstances marked by duress, or clear inequality between the parties, or where the conditions upon which such a contract was based have fundamentally changed, causing thereby unforeseen major distortions in the relations between the parties and thus rendering the contract unfair or oppressive to either of the parties. Aiming at ensuring fairness to all parties concerned, review or renegotiation in such situations should be undertaken in accordance with applicable legal principles and generally recognised legal practices.*[75]

While, after a far-reaching analysis, Berger reaches no firm conclusions, he does posit that there are "good reasons to assume the existence of such a transnational legal principle".[76]

This being said, and as mentioned briefly above (and further below), the Supreme Court of Canada has very recently held that the duty to act in good faith cannot give rise to an obligation to renegotiate a contractual provision that had been duly negotiated by the parties.[77] The Court stressed upon the fact that "good faith serves to protect the equilibrium of a contract, it cannot be used to violate that equilibrium and impose a new bargain on the parties".[78]

---

74 This school of thought asserts that the true agreement of the parties foresaw a quid pro quo consistent with the agreement's equilibrium at the time of contracting, and hence any modification is in reality the enforcement of the original agreement.

75 UN-ECOSOC 1979 Transnational Corporations: Codes of Conduct, Formulations by the Chairman, Art 5 reprinted in Norbert Horn (ed), *Legal Problems of Codes of Conduct for Multinational Enterprises* (1980) 493 and 494. (Even here, however, there is a condition that the contract's applicable law should permit such renegotiation, which is a recurring theme.) Other international law sources include Article 62 of the 1969 Vienna Convention on the Law of Treaties: "A fundamental change of circumstances which has occurred with regard to those existing at the time of the conclusion of a treaty, and which was not foreseen by the parties, may not be invoked as a ground for terminating or withdrawing from the treaty unless: the existence of those circumstances constituted an essential basis of the consent of the parties to be bound by the treaty; and the effect of the change is radically to transform the extent of the obligations still to be performed under the treaty." Houtte also points out, however, that this remains subordinate to the overriding principle of *pacta sunt servanda* (see fn 58, pp112 and, citing other authorities, 121). (One might also note that Article 62 speaks only of termination and withdrawal.)

76 Berger (see fn 70).

77 *Churchill Falls (Labrador) Corp v Hydro-Québec*, 2018 SCC 46; see J Bailey, M Turrini and L Robottom, Good Faith and Renegotiation, White & Case LLP client alert, 20 November 2018.

78 *Churchill Falls (Labrador) Corp v Hydro-Québec*, 2018 SCC 46, p7.

A short review of the recent law in civil law jurisdictions which have traditionally been known not to embrace the notion of a general right of hardship shows that the debate is losing much of its appeal.

4.3 Growing acceptance in (a couple of) civil law countries

(a) Quebec

The Supreme Court of Canada underlined in a 2018 decision, after analysing the principle of hardship (or 'doctrine of unforeseeability') as set out in Article 6.2.2 of UNIDROIT principles and the theory of *imprévision* newly adopted under French legislation, that such doctrine was not recognised under Quebec civil law.[79] The Court went on to indicate that the Quebec courts have generally been reluctant to develop their own law on unforeseeability due to the "political and social nature of the considerations underlying the choice to incorporate into the general law a rule that requires the renegotiation of a contract following an unforeseeable event".[80]

Thus, while there may be scope – in the civil law context, at least – to argue for a more general obligation, such a position is not universal.

(b) Belgium

The Belgian law of contracts used to align with the old French position, with both regimes having the Civil Code of 1804 as their basis. The Belgian Court of Cassation had taken the same position in 1921 as its French counterpart had in the *Canal de Craponne* case,[81] by refusing to allow the court to revise the conditions of performance of a contract in cases of disruption of economic circumstances. A 2009 decision of the Belgian Court of Cassation – which had hitherto denied the existence of such a principle in Belgian law – surprised a number of commentators.[82] The court ruled that changed circumstances which were not reasonably foreseeable at the time of the contract and which make performance more onerous may in certain cases constitute an impediment under Article 79 of the Convention on Contracts for the International Sale of Goods (CISG) excusing non-performance.[83] This had therefore led some commentators to ask whether Belgium would move towards a more general principle of hardship over and beyond the CISG. Similarly, the fact that other legal mechanisms have been used indirectly by the courts to adapt the contract

---

79 *Churchill Falls (Labrador) Corp v Hydro-Québec*, 2018 SCC 46.

80 *ibid*, para 96.

81 See section 4.3(c) below.

82 Cass (Belge) June 19 2009, C.07.0289.N, cited at Dewez et al, "The Duty to Renegotiate an International Sales Contract Under CISG in Case of Hardship and the Use of the Unidroit Principles", *European Review of Private Law* 1-2011, p101.

83 Article 79 of the United Nations Convention on Contracts for the International Sale of Goods, 11 April 1980, 1489 UNTS 3, 19 ILM 671, available at www.cisg.law.pace.edu/cisg/text/treaty.html.

in cases of an unforeseen change of circumstances – for instance, under the cover of interpretation – has given the impression that the door would be open for a recognition of substantial change of circumstances in Belgian domestic law.[84]

Finally, following the reform of the French law of contracts in 2016, the Belgian government rapidly launched a similar reform of the Belgian law of obligations. A bill was recently brought before the House of Representatives[85] and even though there is no clear indication of if and when the new law will be enacted, the chances of adoption seem to be high.[86] As was the case with France in 2016, Belgium plans to have a provision in its revised Civil Code allowing for adaptation or termination of a contract in case of a substantial change of circumstances. The proposed Article 77 mirrors the new French provision on imprévision but also provides some guidance for the judge – he may adapt the contract "with a view to making it conform to what the parties would have reasonably agreed upon when the contract was entered into if they had taken the change of circumstances into account".

(c) *France*

In its seminal decision in the *Canal de Craponne* case in which currency depreciation over three centuries had rendered the contractual price derisory and out of all proportion to the counterpart,[87] the Court of Cassation – the French supreme court – had rejected the notion that in cases of hardship, any duty to renegotiate, let alone a judicial adaptation of the contract, existed outside the realm of French administrative law.[88]

However, in a departure from the legal position firmly established by the *Canal de Craponne* decision, the 2016 reform on contract law[89] has introduced the theory of *imprévision* into the French Civil Code. The new Article 1195 now provides for a multi-phased procedure in the event of an unforeseen change of circumstances, the risk of which had not been accepted by the parties and which renders performance excessively onerous. First, the parties go through an amicable phase of renegotiation of the contract, during which all contractual obligations remain valid. Secondly, *in the event of refusal of the other party to renegotiate or in the event that the renegotiation is not successful, the parties may agree to terminate the contract, or they may mutually request the judge to adapt the contract.*

---

84 See Civ Brussels, 24 May 1884, Pas, 1884, III, p247 cited in E Hondius and C Grigoleit, *Unexpected Circumstances in European Contract Law* (Cambridge University Press, 2011) 161.

85 NautaDutilh, "Reform of the Belgian Civil Code New Bill on the Law of Evidence: what does it mean for your business?", 21 November 2018.

86 M Fontaine, "The Evolution of the Rules on Hardship" in F Bortolotti and D Ufot, *Hardship and Force Majeure in International Commercial Contracts* (International Chamber of Commerce, 2018) 31.

87 French Court of de Cassation, Arrêt du Canal de Craponne of 6 March 1876, Dalloz 1876, at 193.

88 Query, of course, the application of public international law principles to commercial agreements of the type encountered in the energy industry.

89 Ordinance No 2016-131 of 10 February 2016 on the reform of contract law, the general regime and regime on the proof of obligations.

Thirdly, absent an agreement within a reasonable time period, the judge may, at the request of either party, adapt the contract or terminate it on the date and on the conditions determined by the judge.[90]

Does it entail then that the parties would be obliged to attempt to renegotiate contractual terms before being able to seek court intervention? It appears not necessarily, where one party does not wish to play ball; such intervention is subject not only to renegotiation being unsuccessful but also to refusal to renegotiate.[91] However, some commentators[92] suggest that, where the court is asked to adapt or terminate the contract following a mere refusal by a party to renegotiate, that party should be able to explain why he had refused to renegotiate or otherwise face the risk of being held liable for a breach of good faith.[93]

Even before the recent reform of French legislation, the principle of good faith was already being invoked to impose a duty on parties to negotiate. For instance, in 2005, an ICC tribunal applying French law invoked, among other things, the provision of the French Civil Code on good faith, to order negotiation rather than termination of an agreement due to hardship. The tribunal found that the general principle of good faith in contract performance contained in the Code implicitly imposes a duty on parties to negotiate in the face of changed circumstances so that performance of the contract does not cause excessive hardship, especially in long-term contracts.[94]

Article 1195 does not apply to obligations resulting from operations on financial securities, owing to the naturally inherent risks of such products.[95] *More generally, however, the introduction of the theory of imprévision in the French Civil Code, and in particular, the possibility for a party to unilaterally ask the court to adapt or terminate the contract, has been enacted despite criticism from many fronts, with French authors and practitioners fearing for the parties' autonomy and the binding force of the contract*[96] *– so much so that the government had to clarify that contractual parties are free to set aside the application of Article 1195 and/or declare that they assume the risk inherent to an unforeseen change of circumstances rendering performance of the contract excessively onerous.*[97] *This is therefore different from those countries, such as many in Northern Africa, where stipulations seeking to limit the notion of hardship are void.*

---

90 See Annex 1, section 1 below for a full citation of Article 1195.

91 G Debernardi, "Imprévision: de l'expérience italienne à l'innovation française", Les Petites Affiches 23 March 2018, n° 134q0, p7.

92 G Debernardi, "Imprévision: de l'expérience italienne à l'innovation française", Les Petites Affiches 23 March 2018, n° 134q0, p7.

93 Good faith is embedded in the introductory provisions of French contract law. See Article 1104, under Chapter I (I) "Introductory Provisions – The Contract" of the new Civil Code.

94 ICC Case 9994, Excerpt in *ICC International Court of Arbitration Bulletin* (Special Supplement, 2005) 79–80.

95 Article L211-40-1 of the French Monetary and Financial Code.

96 H Le Nabasque, L'imprévision revisitée, Bulletin Joly Sociétés n°01, January 2018, p1.

97 Report to the President of the Republic relating to Ordinance No. 2016-131 of 10 February 2016 on the reform of contract law, the general regime and regime on the proof of obligations, JORF No. 0035 du 11 February 2016, text n° 25.

Comparative studies in the past tended to identify France and Belgium as countries isolated in their strict attachment to the binding force of the contract.[98] With the new legislation in one and imminent reform in the other, it may now be possible to argue for a more general obligation in the civil law context as far as hardship is concerned (at least in continental Europe).

5. A role for national laws?

So where does this discussion leave us? To the extent that one can speak of general notions of hardship, what role might they play in the context of long-term gas supply agreements? One may consider both cases where a price reopener has been agreed by the parties (will national hardship provisions still operate?) and those where no reopener has been agreed.

5.1 Issues that a hardship claim faces

In considering the utility of this doctrine, one can see the issues that a party claiming hardship under a gas sales agreement may face:

- The hurdle over which the claimant must leap in order to activate the trigger can be high. One finds that in many states, one speaks of "excessively onerous" standards or events that "threaten the debtor with exorbitant loss".[99]
- Some clauses require that the claimant first seek termination before a refusal may trigger a renegotiation. It should be readily apparent that purchasers under many long-term arrangements may think long and hard before risking a situation where a supplier agrees to terminate.
- Some provisions require that the underlying event be of a general nature, thereby arguably omitting cases where the problems are specific to a given claimant (although this notion of generality is also often an implied, sometimes express, requirement under certain price reopeners).[100]
- There is also uncertainty in many cases as to the scope of the underlying enquiry into the claimant's position. Is the investigation subjective or objective? And in this light, is one looking at the situation of the party under the contract in question, or at the overall position of that party (for which losing the contract may not be the end of the world)? Can one go even further and look at the group of which the claimant company forms a part?
- In terms of result, hardship clauses rarely, if ever, provide a roadmap for a tribunal to follow in resolving any impasse (whereas price reopeners

98 P Wery, *Droit des Obligations, I*, 2nd edn (Larcier, Brussels, 2011), paras 580–83 cited in in F Bortolotti and D Ufot, *Hardship and Force Majeure in International Commercial Contracts* (International Chamber of Commerce, 2018) 31.

99 See French Civil Code, Article 1195 and Italian Civil Code, Article 1467 at Annex 1, section 1 below, and Civil Code of Jordan, Article 205, Law 43/1976 at Annex 1, section 5 below.

100 However, see Hungarian law, discussed in section 3.1(a) above.

sometimes do provide some idea as to where the parties may tread). The risk is that the tribunal may be wary of proceeding without a more substantial contractual arrangement, or conversely could rewrite the agreement in a way that neither party intended.

- In the same vein, hardship provisions can be said generally to seek to establish contract 'equilibrium', which appears to involve an objective analysis, rather than any broader notion of 'fairness' (a more subjective analysis). For some commentators, this is seen as unnecessarily restrictive.[101] Price reopeners are often tied to the stricter notion of marketability.
- The aims of the two different provisions can also differ markedly. In the simplest terms, hardship is often seen as a means of sharing the pain, which in no way ensures that a buyer will cover its costs or even retain a margin in a hardship situation. This can be compared with many price reopeners, which look to the preservation of the marketability of the gas in question (although even here, these questions are the subject of serious debate). In 2015, as an example, an ICC tribunal ruled that a price revision clause providing for the buyer's ability to economically market gas (the "in any case clause") should be read as providing a safety net and protecting the buyer from unsustainable losses, and not as an implication that he is entitled to anything more than a reasonable margin.[102]

5.2 Is there nevertheless a role for hardship?

So does this mean that there is little or no role for hardship principles in the area of long-term gas sales?

It would appear that certain tribunals believe so, especially in the presence of a clear price revision clause. In a 2013 ICC Case, the respondent argued that the revision clause invoked by the claimant to request an extraordinary revision of the contract price was by its nature a hardship clause, so that the contractual condition of upheaval in the economic and/or energy and gas market conditions needed to be unpredictable. The tribunal decided that the will of the parties was to establish a revision clause triggered by conditions less drastic than those of hardship and that UNIDROIT Principles apply only if the parties have expressly referred to them in their contract.[103]

---

101 Wolfgang Peter, Jean-Quentin de Kuyper, Bénédict de Candolle, *Arbitration and renegotiation of international investment agreements* (Kluwer Law International, 1995) 2.

102 See ICC Case 19374 cited in G Block, "Arbitration and Price Renegotiation in Energy Contracts" (2017) 2 *ICC Dispute Resolution Bulletin* 55.

103 See ICC Case 18123 cited in G Block, "Arbitration and Price Renegotiation in Energy Contracts" (2017) 2 *ICC Dispute Resolution Bulletin* 53. The tribunal notably stated that: "Understandable though it may be that the revision imposed [by hardship] outside any contractual agreement is the ultimate solution to avoid the contractor being overwhelmed by the burden of his obligations, the rebalancing agreed in advance by the contracting parties must be done under the agreed conditions."

Even in contracts in which a revision clause is complemented with a hardship clause, arbitral tribunals tend to consider hardship as the fall-back, not needing to be examined if price adjustment is granted under the price revision clause.[104]

It may, however, be premature to be so categorical as to rule out the role of hardship completely:

- Given that the bulk of gas price reopener clauses look (partially or completely) at events in and affecting the buyer's end-user market alone,[105] there may well be cases where the relevant events arise in areas such as the supply chain, which have nothing to do with the end market.
- Certain 'non-price' issues (eg, quantities or quality) can give rise to substantial problems but may not be covered by a given reopener clause.[106]
- There may be cases (albeit few) where a gas price reopener gives rise to an obligation to negotiate, but the clause does not provide the arbitrators with the power and authority to impose a solution above and beyond a finding on whether that obligation to negotiate was fulfilled.[107] Here again, hardship may be the only avenue open.
- A number of price reopener clauses contain detailed provisions as to the review period to which the parties may refer; hardship provisions are often much more liberal in terms of the period over which the hardship analysis is to be undertaken.

Thus, while for a number of commentators, the role for hardship appears limited – particularly in the face of a well-drafted gas price clause[108] – this does not mean that there is no such role. Case law confirms that the two can coexist, and indeed that hardship can come into play where a gas price clause exists and is applied.[109] Some legal systems, such as those of North Africa and the Middle

104 See ICC Case 17648 cited in G Block, "Arbitration and Price Renegotiation in Energy Contracts" (2017) 2 *ICC Dispute Resolution Bulletin* 53.

105 Many (not all) clauses concentrate solely on events in the buyer's market and/or affecting the marketability of the gas in that market.

106 In the writer's experience, this can vary from case to case.

107 See eg, KP Berger, above fn 70 at pp1367–68, citing *Kuwait v Am Indep Oil Co*, 21 ILM at 1004 ("An obligation to negotiate is not an obligation to agree"); *Wintershall AG v Gov't of Qatar*, Ad hoc tribunal, 28 ILM 795 at 814, 841("It is clear that such a duty [to negotiate] does not include an obligation ... to reach agreement ... [and] the Government [is not] legally required to enter into such an agreement, however reasonable it may be").

108 G Block, "Arbitration and Price Renegotiation in Energy Contracts" (2017) 2 *ICC Dispute Resolution Bulletin* 52.

109 See discussion by G Block of an award where the indexation clause of a contract, which indexed the price of gas supplied to an electricity producer 100% to the price of coal, caused a significant imbalance when oil prices and electricity prices soared. This brought into play the hardship clause to restore the equilibrium; G Block, "Arbitration and Changes in Energy Prices: A Review of ICC Awards with respect to Force Majeure, Indexation, Adaptation, Hardship and Take-or-Pay Clauses" (2009) 20(2) *ICC International Court of Arbitration Bulletin* 51, p56. One can equally see arguments in favor of applying hardship in circumstances where a price clause is found inadequate or inapplicable.

East, in fact insist upon a role for hardship regardless of what the parties might say.

It may also be the case that contract drafters themselves may want to provide for a hardship clause in order to cover matters which are not the subject of traditional gas price reopener clauses. In fact, the main message conveyed during the recent ICC Conference on Hardship and *Force Majeure* in International Commercial Contracts[110] was that negotiators should first systematically verify the rules on change of circumstances in the applicable law, and then depart from such legal provisions in case they are not satisfied.[111]

The possibility of raising a hardship claim thus should not be overlooked.

Annex 1: National provisions on hardship[112]

1. Europe

1.1 Italy

Article 1467 of the Civil Code:

In contracts with continuous or periodical execution or adjourned execution and in case that the obligation of one of the parties has become excessively onerous due to extraordinary and unpredictable events, the party who is obliged to such performance can demand the dissolution of the contract with the effects laid down in Art 1458. The dissolution cannot be demanded if the supervening burdensome event is part of the normal risk of the contract. The party against which the dissolution is demanded can prevent this by offering to modify equitably the conditions of the contract.

Article 1468:

In the case contemplated in the preceding Article, if the contract is one in which only one of the parties has assumed obligations, he can demand a reduction in his performance or a modification of the manner of performance, sufficient to restore it to an equitable basis.

1.2 Netherlands

Article 6:258 of the Civil Code:

(1) The Court may, upon request of one of the parties, modify the effects of an agreement or terminate it in part or in its entirety on the basis of unforeseen circumstances of such nature that the other party may not, according to the

110 ICC Conference on Hardship and Force Majeure in International Commercial Contracts, Paris, International Chamber of Commerce, 29 November 2018.

111 See M Fontaine, "The Evolution of the Rules on Hardship" in F Bortolotti and D Ufot, *Hardship and Force Majeure in International Commercial Contracts* (International Chamber of Commerce, 2018) 37.

112 Free translations. The writer cannot vouch for the accuracy of all translations; in case of particular interest, the reader should have the relevant translation verified.

criteria of reasonableness and fairness, expect the agreement to be maintained in an unmodified form. The modification or termination may be given retroactive effect.

(2) The modification or termination may not be pronounced to the extent that these circumstances are to be borne by the party relying upon them, according to the nature of the agreement or commercial common opinion.

(3) For the purpose of this Article, a person to whom a right or obligation from the agreement has passed, is equated with an original party to that agreement.

1.3 France

Article 1195 of the new Civil Code:[113]

If a change of circumstances that was unforeseeable at the time of the conclusion of the contract renders performance excessively onerous for a party who had not accepted the risk of such a change, that party may ask the other contracting party to renegotiate the contract. The first party must continue to perform his obligations during renegotiation.

In the case of refusal or the failure of renegotiations, the parties may agree to terminate the contract from the date and on the conditions which they determine, or by a common agreement ask the court to set about its adaptation. In the absence of an agreement within a reasonable time, the court may, on the request of a party, revise the contract or put an end to it, from a date and subject to such conditions as it shall determine.

This provision does not apply to obligations resulting from operations on financial securities and contracts[114] *and parties are free to set aside the application of Article 1195 and/or declare that they assume the risk inherent to an unforeseen change of circumstances rendering performance of the contract excessively onerous.*[115]

1.4 Germany

Section 313 of the Civil Code:

Interference with the basis of the transaction

(1) If circumstances which became the basis of a contract have significantly changed since the contract was entered into and if the parties would not have entered into the contract or would have entered into it with different contents if they had foreseen this change, adaptation of the contract may be demanded to the extent that, taking account of all the circumstances of the specific case, in particular the contractual or statutory distribution of risk, one of the parties cannot reasonably be expected to uphold the contract without alteration.

(2) It is equivalent to a change of circumstances if material conceptions that have become the basis of the contract are found to be incorrect.

---

113 As applicable as from 1 October 2016.

114 Article L211-40-1 of the French Monetary and Financial Code.

115 Ordinance No. 2016-131 of 10 February 2016 on the reform of contract law, the general regime and regime on the proof of obligations.

(3) If adaptation of the contract is not possible or one party cannot reasonably be expected to accept it, the disadvantaged party may revoke the contract. In the case of continuing obligations, the right to terminate takes the place of the right to revoke.

1.5 Spain[116]

All contracts are binding unless there is a change of circumstances. The question of unexpected circumstances has been approached as an implied *rebus sic stantibus* clause. The Spanish courts have developed a doctrine of *rebus sic stantibus* alongside Article 1091 of the Civil Code.[117] In the event of changed circumstances, this doctrine permits a party that is unduly burdened to be discharged from the contract or to request the adaptation of the contract to the new situation.

The conditions stipulated by the courts for the application of the *rebus sic stantibus* doctrine are strict:

- The contract is a long-term contract or a contract in which at least one of the parties has not yet performed.
- The basis of the contract has been altered. This is accepted if:
  - the contract has become excessively burdensome for one of the parties; or
  - the purpose of the contract has been totally frustrated.
- The change of circumstances is extraordinary in nature and unforeseen.
- Neither party has accepted the risk of the change of circumstances as a contractual obligation.
- The person invoking the change of circumstances is not responsible for this change.

The application of this doctrine can result in either adaptation of the contract, restoring the equilibrium, or termination of the contract. The courts prefer to adapt the contract. Therefore, termination of the contract is granted only when it is not possible to restore the balance between the performances.

1.6 Portugal[118]

Article 437 of the Civil Code:

[I]f the circumstances on which the parties based their decision to contract change unexpectedly, the burdened party has a right to terminate the contract, or to alter

---

116 This involves private law; see Ugo Draetta, Ralph B Lake and Ved P Nanda, *Breach and Adaptation of International Contracts: An Introduction to Lex Mercatoria* (Hondius and Grigoleit, 1992) (see fn 84), pp127–28. For public law, see M Franch and J Torrelles, "Spain" in Rozen Noguellou & Ulrich Stelkens (eds), *Droit comparé des Contrats Publics/Comparative Law on Public Contracts* (2010).

117 Article 1091 states: "Obligations arising from contracts are legally binding on the contracting parties, and should be fulfilled accordingly."

118 See Hondius and Grigoleit (see fn 84), p135.

it according to fair judgment, as long as the requirement of the obligations which it assumed seriously affects the principles of good faith and is covered by the actual risks inherent in the contract.

1.7 Greece[119]

Section 388 of the Civil Code:

Unforeseeable change of circumstances. If having regard to the requirements of good faith and business usages the circumstances on which the parties had based the conclusion of a bilateral agreement have subsequently changed on exceptional grounds that could not have been foreseen and the performance due by the debtor taking also into consideration the counter-performance has as a result of the change become excessively onerous the Court may at the request of the debtor and according to its appreciation reduce the debtor's performance to the appropriate extent or decide the dissolution of the contract in whole or with regard to its non performed part.

If the dissolution of the contract has been decided the obligations to perform arising therefrom shall be extinguished and the contracting parties shall be reciprocally obligated to make restitution for the performances by which each benefited pursuant to the provisions governing enrichment without just cause.

1.8 Switzerland

Hans van Houtte notes the following:

The Swiss Federal Tribunal has admitted that some long-term contracts may be terminated because of an unforeseeable and fundamental change of circumstances on the basis of Article 2 of the Code civil (good faith). Only changes which would unjustly enrich one of the parties give rise to such relief. Rebus sic stantibus, however, has to be applied restrictively.[120]

2. Nordic countries

2.1 Sweden

Article 36 of the Swedish Contract Act:

A contract term or condition may be modified or set aside if such term is unreasonable with regard to the contents of the agreement, the circumstances at the time the agreement was entered into, subsequent circumstances and circumstances in general. Where a term is of such importance to the agreement that it would be unreasonable to demand the continued enforceability of the remainder of the

---

119 See Draetta, Lake and Nanda (see fn 116); Hondius and Grigoleit (see fn 84), pp138–44.

120 H van Houtte (see fn 58), p113. He cites the following authority: Federal Tribunal, 10 October 1933, ATF (59), 1933, II, 372. See moreover Federal Tribunal, 13 January 1975, ATF (101), 1975, II, 17, 19; Federal Tribunal, 26 September 1974, ATF (100), 1974, II, 345; Federal Tribunal, 7 December 1971, ATF (97), 1971, II, 390; Federal Tribunal, 18 September 1981, ATF (107), 1981, II, 343.

agreement with its terms unchanged, the agreement may be modified in other respects or may be set aside in its entirety.
Upon determination of the applicability of the provisions of the first paragraph, special consideration shall be paid to the need for protection of the party which, as a consumer or otherwise, is in a position of inferiority in the contractual relationship.

See also Article 36 of the Finnish Contracts Act and Article 36 of the Danish Contracts Act.

3. South America

3.1 Brazil

Article 478 of the Civil Code:

For contracts of continuing or deferred performance, if the performance by one of the parties becomes excessively onerous, providing an undue advantage to the other, and as a result of extraordinary and unforeseen events, the disadvantaged party may request the termination of the contract. The effects of a judgment of termination shall be retroactive to the date of the summons.

Article 479 of the Civil Code:
Termination can be avoided if the defendant agrees to modify equitably the terms of the contract.

Article 480 of the Civil Code:
If, in a contract, the obligations of performance apply only to one of the parties, such party may request that its obligations be reduced or modified as to its performance in order to avoid excessive burden.

3.2 Argentina

Article 1198 of the Civil Code:

Contracts are to be made, interpreted and executed in good faith, according to what the parties understood or could have understood, acting with care and foreseeability.
If, in bilateral contracts with deferred or continued performance, the performance of one of the parties becomes excessively onerous because of extraordinary and unforeseeable events, the prejudiced party may demand the termination of the contract. The same principle shall apply to aleatory contracts when excessive onerousness results from causes extraneous to the very risk of a contract.
In contracts of continued performance termination shall not affect that which has already been performed.
Termination shall not be granted if the prejudiced party has been at fault or in delay.

The other party may prevent termination by offering to improve the effects of the contract in an equitable fashion.

3.3 Peru

Article 1440 of the Civil Code:

Definition. In commutative contracts of continuous, periodic, or deferred performance, if performance becomes excessively onerous due to extraordinary or unforeseeable events, the aggrieved party may request a court to reduce or increase the consideration, such that the excessive burden will cease.

If this were not possible due to the nature of the performance, the circumstances or if the defendant so requests, the court will order the termination of the contract. The termination shall not extend to the obligations already performed.

Article 1441:

Scope of the excessive burden of the obligation. The provisions contained in Article 1440 apply:

1. *To commutative contracts of immediate performance, when performance by one of the parties has been deferred due to a cause not attributable to such party.*
2. *To contracts of chance, when the excessive burden arises from causes that are extraneous to the risk inherent to the contract.*

Article 1442:

Excessive burden in contracts providing for performance by one party. Where only one of the parties undertook obligations under a contract, such party shall be entitled judicially to request the reduction of the obligation such that the excessive burden will cease.

If the obligation cannot be reduced, the second paragraph of Article 1440 shall apply.

Article 1443:

Inadmissibility of the action for excessive burden. The action for excessive burden of the obligation shall not proceed when its performance is deferred due to fraud or fault of the aggrieved party.

Article 1444:

Nullity of waiver of the action. Waiver of the action for excessive burden of the obligation shall be null and void.

Article 1445:

Expiration of the action. The action for excessive burden of the obligation shall expire upon three months of the occurrence of the extraordinary and unforeseeable events referenced in Article 1440.

Article 1446:

Term of expiration. The initial term of expiration referenced in Article 1445 shall run as of the moment when the extraordinary and unforeseeable events disappear.

4. North Africa

4.1 Algeria

Article 107(3) of the Civil Code:

When ... as a result of exceptional and unforeseeable events of a general character, the performance of the contractual obligation, without becoming impossible, becomes excessively onerous in such a way as to threaten the debtor with exorbitant loss, the judge may, according to the circumstances, and after taking into consideration the interests of both parties, reduce to reasonable limits, the obligation that has become excessive. Any agreement to the contrary is void.

4.2 Libya[121]

Article 147 of the Civil Code:

1. *The contract makes the law of the parties. It can be revoked or altered only by mutual consent of the parties or for reasons provided by for the law.*
2. *When, however, as a result of exceptional and unpredictable events of a general character, the performance of the contractual obligation, without becoming impossible, becomes excessively onerous in such a way as to threaten the debtor with exorbitant loss, the Judge may, according to the circumstances and after taking into consideration the interests of both parties, reduce to reasonable limits the obligation that has become excessive. Any agreement to the contrary is void.*

Article 657:

1. *When a contract is concluded on a lump sum basis according to a design agreed with the employer, the contractor has no claim to an increase of price, even if modifications and additions are made to the design, unless such modifications or additions are due to the fault of the employer or have been authorised by the employer and the price thereof agreed with the contractor.*
2. *Such agreement should be made in writing unless the principal contract was concluded verbally.*
3. *The contractor has no claim to an increase of price on the grounds of an increase in the price of raw materials, labour or any other item of expenditure, even if such increase is so great as to render the performance of the contract onerous.*
4. *When, however, as a result of exceptional events of a general character which could not be foreseen at the time the contract was concluded, the economic*

121 See TJ McDonald and OA Tarhuni, *Compensation for Payment Delays under Libyan Laws, The International Construction Law Review*, October 1985.

equilibrium between the respective obligations of employer and of the contractor breaks down, and the basis on which the financial estimates of the contract were computed has subsequently disappeared, the Judge may grant an increase of the price or order the resolution of the contract.

Morocco and Tunisia appear not to have adopted the principle of hardship, at least in the sense of its codification.

5. Middle East

5.1 Bahrain

Article 130 of Legislative Decree 19/2001, promulgating the Civil Code of Bahrain:

> *When, however, as a result of exceptional and unpredictable events of a general character, the performance of the contractual obligation, without becoming impossible, becomes excessively onerous in such way as to threaten the debtor with exorbitant loss, the judge may, according to the circumstances, and after taking into consideration the interests of both parties, reduce to reasonable limits by lessening its extent or increasing its consideration, the obligation that has become excessive. Any agreement to the contrary is void.*

5.2 Egypt

Article 147(2) of the Civil Code:

> *The contract makes the law of the parties. It can be revoked or altered only by mutual consent of the parties or for reasons provided for by the law.*
>
> *When, however, as a result of exceptional and unpredictable events of a general character, the performance of the contractual obligation, without becoming impossible, becomes excessively onerous in such way as to threaten the debtor with exorbitant loss, the judge may, according to the circumstances, and after taking into consideration the interests of both parties, reduce to reasonable limits, the obligation that has become excessive. Any agreement to the contrary is void.*[122]

5.3 Iraq

Article 146 of the Civil Code:

1. *When the contract is concluded, it becomes legally binding and can only be revoked or modified in accordance with the law or by the consent of both parties.*
2. *When, however, as a result of exceptional and unforeseeable events of a general character, the performance of the contractual obligation, without becoming impossible, becomes onerous in such a way as to threaten the debtor with exorbitant loss, the court may after taking into consideration the interests of*

---

122 P Fanner and S Marshall, *Translation of the Egyptian Civil Code* (New World Publisher) 29.

both parties, reduce to reasonable limits the obligation that has become excessive if justice so requires. Any agreement to the contrary is void.

5.4 Jordan

Article 205 of the Civil Code:

If as a result of exceptional and unforeseeable events of a general character, the performance of the contractual obligation, without becoming impossible, becomes onerous in such a way as to threaten the debtor with exorbitant loss, the court may, according to the circumstances, and after taking into consideration the interests of both parties, reduce to reasonable limits the obligation that has become excessive if justice so requires. Any agreement to the contrary is void.

5.5 Kuwait

Article 198 of the Civil Code:

Where after and before completion of performance of the contract, general extraordinary circumstances arose which could not have been foreseen on the execution thereof, and which rendered the performance of the obligation arising therefrom oppressive, if not impossible, to the debtor such as will threaten him with heavy loss, the judge may, after weighing the interests of both parties, reduce the oppressive obligation to a reasonable degree either by narrowing its extent or by awarding a balancing interest; any agreement otherwise shall be null.

5.6 Lebanon

In Lebanon, the situation is similar to that under French law, providing for hardship under administrative, but not commercial law.

5.7 Qatar

Article 171 of the Civil Code:

1. *A contract is the law of the contracting parties and so cannot be revoked or modified except with the agreement of the parties or for such reasons as are prescribed by the law*
2. *Notwithstanding, if any general, exceptional events occur that cannot be foreseen as a result of which fulfilment of the contractual obligation becomes, although not impossible, onerous for the debtor such that he is at risk of incurring a substantial loss, the judge may, according to the circumstances and after weighing up the interests of the parties, return such onerous obligation to a reasonable level. Any agreement to the contrary will be void.*

5.8 Syria

Article 148 of the Civil Code:

The contract is the law of the parties. It can be revoked or altered only by the mutual consent of the parties or for reasons provided by the law.

When, however, as a result of exceptional and unpredictable events of a general character, the performance of the contractual obligation, without becoming impossible, becomes excessively onerous in such a way as to threaten the debtor with exorbitant loss, the judge may, according to the circumstances, and after taking into consideration the interests of both parties, reduce to reasonable limits, the obligation that has become excessive. Any agreement to the contrary is void.[123]

5.9 United Arab Emirates

Article 249 of the Civil Code:

When ... as a result of exceptional and unforeseeable events of a general character, the performance of the contractual obligation, without becoming impossible, becomes excessively onerous in such a way as to threaten the debtor with exorbitant loss, the judge may, according to the circumstances, and after taking into consideration the interests of both parties, reduce to reasonable limits, if justice requires so, the obligation that has become excessive. Any agreement to the contrary is void.

5.10 Yemen

Article 211 of the Civil Law:

The contract is legally binding on the contracting parties and it shall not be voided or modified without both parties' consent or for the reasons stipulated by the law, notwithstanding. If as a result of exceptional and unforeseeable events of a general character such as wars and catastrophes, the performance of the contractual obligation, without becoming impossible, becomes onerous in such a way as to threaten the debtor with exorbitant loss, stopping the debtor from proceeding with the contract, and that does not mean an increase or decrease in the prices, the Judge may, according to the circumstances, whether poverty, wealth or others, and after taking into consideration the interests of both parties, reduce the obligation that has become excessive to reasonable limits.

6. Sub-Saharan Africa

6.1 Sudan

Article 117 of the Law of Civil Transactions:

1. *If as a result of exceptional and unforeseeable events of a general character, the performance of the contractual obligation, without becoming impossible, becomes onerous in such a way as to threaten the debtor with exorbitant loss, the court may, according to the circumstances, and after taking into*

123 The Syrian judge would require the debtor to bear all the usual increase, while the unusual increase will be borne equally by the debtor and the creditor (see Cassation 247/80 published in the *Syrian Advocates Bar Journal* (1981) 445).

consideration the interests of both parties, reduce to reasonable limits the obligation that has become excessive. Any agreement to the contrary is void.

2. *The obligation is not considered onerous unless the loss exceeds one-third of the obligation.*

7. Eastern Europe

7.1 Russia

Article 451 of the 1996 Civil Code on "the Amendment and the Cancellation of the Contract Because of an Essential Change of Circumstances":

1. *An essential change of the circumstances, from which the parties have proceeded when concluding the contract, shall be the ground for its amendment or cancellation, unless otherwise stipulated by the contract or following from its substance. The change of the circumstances shall be recognised as essential, if they have changed to such an extent that in case the parties could have wisely envisaged it, the contract would not have been concluded by them or would have been concluded on the essentially different terms.*
2. *If the parties have failed to reach an agreement on bringing the contract into correspondence with the essentially changed circumstances or on its cancellation, the contract may be cancelled, and on the grounds, stipulated by Item 4 of the present Article, it may be amended by the court upon the claim of the interested party in the face of the simultaneous existence of the following conditions:*
   1) *at the moment of concluding the contract, the parties have proceeded from the fact that no such change of the circumstances will take place;*
   2) *the change of the circumstances has been called forth by the causes, which the interested party could not overcome after they have arisen, while displaying the degree of care and circumspection, which have been expected from it by the nature of the contract and by the terms of the circulation;*
   3) *the execution of the contract without amending its provisions would so much upset the balance of the property interests of the parties, corresponding to the contract, and would entail such a loss for the interested party that it would have been to a considerable extent deprived of what it could have counted upon when concluding the contract;*
   4) *neither from the customs of the business turnover, nor from the substance of the contract does it follow that the risk, involved in the change of the circumstances, shall be borne by the interested party.*
3. *In case of the cancellation of the contract because of the essentially changed circumstances, the court shall, upon the claim of any one of the parties, define the consequences of the cancellation of the contract, proceeding from the need to justly distribute the expenses, borne by them in connection with the execution of this contract, between the parties.*

4. *The amendment of the contract in connection with an essential change of the circumstances shall be admitted by the court decision in extraordinary cases, when the cancellation of the contract contradicts the public interests, or if it entails the losses for the parties, considerably exceeding the expenses, necessary for the execution of the contract on the terms, amended by the court.*

Article 709(6) of the 1996 Civil Code provides that a contractor has the right to demand an increase in the price of a contract containing no price escalation clause where there has been a material increase in the cost of materials or in the cost of services rendered to the contractor by a third party.

Article 17 of the Act on Production Sharing Agreements 1995 (as amended July 19 2011) allows for the contract to be amended without requiring a hardship clause should a material change in circumstances take place as defined by the Code.

Administrative contracts in Russia are also subject to provisions on unforeseeable events.

7.2 Lithuania[124]

Article 6.204 of the Civil Code:

1. *If the performance of a contract becomes more onerous for one of the parties, this party shall be bound to perform the contract in accordance with the procedure established in the other Paragraphs of this Article.*
2. *The performance of a contract shall be considered obstructed under such circumstances which fundamentally alter the balance of the contractual obligations, i.e., either the cost of performance has essentially increased, or the value thereof has essentially diminished if: 1) these circumstances occur or become known to the aggrieved party after the conclusion of the contract; 2) these circumstances could not reasonably have been foreseen by the aggrieved party at any time of the conclusion of the contract; 3) these circumstances are beyond the control of the aggrieved party; 4) the risk of occurrence of these circumstances was not assumed by the aggrieved party.*
3. *If the performance of a contract becomes obstructed, the aggrieved party shall have the right to make a request to the other party for the modification of the contract. Such request shall have to be made immediately after the occurrence of the obstructions and the grounds on which the request is based indicated therein. The request for modification of the contract shall not in itself entitle the aggrieved party with the right to suspend performance of the contract. Where within a reasonable time, the parties fail to reach an agreement on the modification of the contractual obligations, any of them may bring an action in court. The court may: 1) dissolve the contract and establish the date and terms*

124 See Hondius and Grigoleit (see fn 84), pp83–84.

of its dissolution; 2) modify the conditions of the contract with a view to restoring the balance of contractual obligation of the parties.

7.3 Slovenia[125]

Articles 112 to 115 of the Slovenian Code of Obligations (2002) provide that if there is a significant change of circumstances, the court can dissolve a contract at the request of one of the parties.

However, adjustment of the contract appears not to be possible unless both parties consent; one party cannot be forced to accept an adjustment.[126]

7.4 Hungary

Section 6:192 of the Civil Code[127] provides that a party is entitled to request to have the contract amended by the court if, in the parties' long-term contractual relationship, performing the contract under the same terms is likely to harm his relevant lawful interests, in consequence of a circumstance that has occurred after the conclusion of the contract. Three cumulative conditions apply: the change of circumstances must have been unforeseen at the time of the conclusion of the contract, must not have been caused by the party seeking amendment of the contract and must arise outside normal business risks.

The court may amend a contract when it is injurious to any substantial rightful interest of one of the parties in consequence of a circumstance arising in the long-term relationship of the parties following the conclusion of the contract.

7.5 Estonia

The new Estonian Law of Obligations Act, in force since 2002 provides, under paragraph 97 on Alteration of Balance of Contractual Obligations:

1. *If the circumstances under which a contract is entered into change after the entry into the contract and this results in a material change in the balance of the obligations of the parties due to which the costs of one party for the performance of an obligation increase significantly or the value of that which is to be received from the other party under the contract decreases significantly, the injured party may demand amendment of the contract from the other party in order to restore the original balance of the obligations.*
2. *Amendment of a contract under the circumstances specified in subsection (1) of this section may be demanded if:*
   1) *at the time of entry into the contract, the injured party could not have reasonably expected that the circumstances might change, and*
   2) *the injured party could not influence the change in the circumstances, and*

---

125 *ibid*, p76.
126 See Slovenian Supreme Court, III Ips 17/93.
127 Act V of 2013 on the Civil Code, 26 February 2013.

3) the risk of a change in the circumstances is not borne by the injured party pursuant to the law or the contract, and

4) the injured party would not have entered into the contract or would have entered into the contract under significantly different terms if the party had known of the change in the circumstances.

3. Amendment of a contract may also be demanded if the circumstances under which the contract was entered into had already changed before the contract was entered into but became known to the injured party after the contract was entered into.

4. The injured party may also demand amendment of the contract retroactively, but not as of a time earlier than the time when the balance of the obligations changed.

5. If the bases for amendment of a contract exist but, due to the circumstances, amendment of the contract pursuant to subsection (1) of this section is not possible or would not be reasonable with respect to the other party, the party aggrieved by alteration of the balance of the obligations may withdraw from the contract or, in the case of a long-term contract, cancel the contract pursuant to the procedure provided for in § 196 of this Act.

8. The Far East

8.1 China

While Chinese contract law does not expressly include a provision for alteration of the contract where there has been such a change in circumstances, Article 26 of "Interpretation II of the Supreme People's Court on Several Issues concerning the Application of the Contract Law of the People's Republic of China" issued by the Supreme People's Court of the People's Republic of China in 2009 provides that a contractual party may request the court to modify or terminate the contract in the event of an unforeseeable, significant change of circumstances which is outside the scope of commercial risks and not caused by force majeure, and where it would be obviously unfair to a party, or the purpose of the contract would be frustrated, if the parties continued to perform the contract. In deciding whether to grant such a request, the court will consider the principle of fairness in light of the circumstances of each case. The Supreme People's Court issued two other policies to further clarify the application of Article 26.

8.2 Japan[128]

There is no statutory law governing changed contractual circumstances in Japan. However, a Japanese court can adapt the contract if it finds unforeseen,

128 See Puelinckx (see fn 11), p47 *et seq.*

substantially changed circumstances (without fault of parties) that cause the obligation of continued performance to violate the principles of good faith and equity.

In his book *The Principle of Changed Circumstances in Civil Law* (1926), Katsumoto Masaaki gave the following definition of 'changed circumstances':

> *The rule states that when the circumstances releasing a legal act, creating a legal relation, change in an unforeseeable way alter the accomplishment of the act and before expiry of its consequences, due to the influence of facts not attributable to the parties, with the result that the creation of legal consequences in the original sense, or their continuation, is to be considered as unreasonable in the light of the rule of equity and good faith, then such legal consequences are to be adapted in equity and good faith.*[129]

This definition became traditional in Japan.

The Japanese Supreme Court adopted this doctrine in 1944, but has been very restrictive in its application. However, the lower courts were more lenient in applying it.[130]

The author would like to thank Samy Markbaoui, Hinda Rabkin and Yutty Ramen for their assistance in the preparation of this chapter, as well as Pedro and Filipa at PLMJ. This chapter represents his personal opinions and as such should in no way be attributed to partners or colleagues at White & Case LLP.

129 *ibid*, pp62–63.

130 *ibid*; a typical application of this is as follows: "The parties agreed in 1935 that a buyer would have an option to buy a piece of land for a set price, such option to last twenty years. When the buyer asked to exercise its option, in 1955, the actual value of the land was estimated at 620 times the price fixed by the contract, as the result of the enormous inflation and soaring land prices following Japan's defeat in World War II. The change in circumstances meant that upholding the contract would violate principles of good faith and equity."

Procedural issues in gas price review arbitrations

Oliver Browne
Philip Clifford QC
Latham & Watkins

1. Introduction

Whilst gas (including LNG) price review arbitrations take different forms, they make up a discrete category of arbitration conducted between a relatively limited number of potential participants. Typically involving very high-value disputes under medium to long-term sale and purchase contracts, they are often heavily focused on expert evidence of the relevant gas, LNG and/or energy markets, which can involve highly confidential data. They also differ from most other arbitrations in that they concern potential changes being made by the arbitral tribunal to the underlying contract (by way of adjustment to the price formula), rather than simply addressing questions of liability for breach and compensation for loss and damage.

Price review provisions are found primarily in medium to long-term contracts, where the ability to adjust the price over time is important, and the parties are focused on their ongoing relationships. As a result, they often provide for complex qualifying criteria and extensive procedural steps aimed at limiting disputes and settling issues without the need for arbitration. However, as with any qualifying criteria and other multi-tiered dispute resolution mechanisms, these can themselves (perhaps counter-intuitively) give rise to problems.

Indeed, most of the distinguishing characteristics of gas price review arbitrations can give rise to procedural issues, and this chapter addresses the most common that occur. This chapter first looks at the procedural issues related to pre-arbitration steps and then considers the issues that often arise during the arbitration itself. However, as will be apparent, issues can have knock on effects, so that issues arising at a pre-arbitration stage can easily spill over into the arbitration itself.

Price review provisions can take many different forms, so the potential issues are discussed in general terms and by reference to contract provisions that are commonly encountered. However, any given case must, of course, be considered by reference to its particular facts and legal framework (including relevant substantive and procedural laws).

2. Procedural issues related to pre-arbitration steps

2.1 Trigger event(s), Reference Periods and Review Notices

The process leading to a price review often begins with a 'trigger event(s)', being the event(s) set out by the parties in their contract as giving rise to the right to a price review. There are many different types of trigger event, which can be as simple as a specified date being reached, sometimes upon the expiry of a period of time following a previous price review date. Such simple triggers have the advantage of giving rise to few procedural issues.

However, there is little point in having a price review if there is no justification for adjusting the price. It is therefore common for trigger events to include references to the matters that will determine whether there is in fact a need to make an adjustment. For example, these may be changes in market circumstances, together with a timing requirement (eg, three years having passed since the last price review date), which aims to spread out price reviews over the term of the contract. There is sometimes also an option within the price review to request a special price review(s), at any time, to cater for any urgent need for a review (eg, in response to a specific, significant, event).

The potential procedural difficulty with the more complex triggers, which themselves require at least the basis of a justification for adjusting the price, is that they can make establishing the jurisdiction of any future arbitral tribunal, the body that may ultimately have to decide if the triggers have been satisfied, more difficult. This is because complex triggers may give rise to questions as to whether they have been satisfied at all and, if not, whether a right to arbitrate has in fact arisen. However, whilst some parties may raise jurisdictional issues in this regard, in practice tribunals generally find themselves able to determine their own jurisdiction. Where it is not appropriate to deal with questions of jurisdiction at an early stage, tribunals may also choose to defer determination of those questions until their final award. By doing this, tribunals can deal with their jurisdiction in light of their view on the merits.

The trigger events must take place and/or have an effect satisfying any prescribed criteria within the relevant period (often referred to as the 'Reference Period' or the 'Review Period'), which is generally a period between the dates when price reviews can be requested. The assessment of the effect of the trigger events, which is usually the basis for any adjustment to the price formula, often compares the situation at the start of the Reference Period with that at the end. Questions frequently arise as to the period and duration of the relevant Reference Period as a result of ambiguities in the contractual language that tribunals may only be able to address in any final award.

Where a party considers that there has been a qualifying trigger event(s) and wishes to commence the price review process, it may be required to send a

Review Notice to the other party. There are sometimes complex requirements for the timing and form of such notices, but the detail required (if any) varies.

Given the limited availability of price reviews during the life of the contract, there are often contractual restrictions as to when a Review Notice can be given. For example, a Review Notice may be required to be served within a designated period before the effective date for any price formula adjustment. Where the contract is not clear, and, for example, does not specify whether the Review Notice can be served on the start or end dates for that period, or when service is to be effective, this can lead to issues.

In terms of its required content, a Review Notice can sometimes simply be a notification that the contract process for a price review should be started, but in other cases it can be a more detailed explanation of what has happened, the consequences and the adjustment sought (and, of course, it could be anything in between).

The required timing and content of Review Notices are essentially matters requiring an interpretation of the parties' contract, and where the drafting is incomplete or unclear, procedural issues are likely to be raised.

It is also in the nature of price reviews that opposing parties may simultaneously seek such a review contending that opposite adjustments be made to the price formula. This can give rise to procedural issues where this issue is not properly addressed in the price review provisions (which may, for example, only anticipate one party requesting a review at any one time, or be unclear in this regard).

The various contractual provisions concerned with the commencement of a price review demand a careful focus on both substance and procedure from the start. Disputes can arise over, for example, the timing and/or content of Review Notices where these are not clearly addressed. It is generally a matter of contractual interpretation whether non-compliance with any given provision affects the validity of the price review, the admissibility of any claim and/or the jurisdiction of any tribunal subsequently appointed. However, where the position is left unclear in the parties' contract, this can be fertile ground for issues to arise.

2.2 Negotiation between the parties

It is common for contracts to provide that, in the event that a Review Notice is served, the parties are to seek to resolve the matter through negotiation. In some circumstances, events or market conditions may have led the parties to discuss the situation well in advance of any Review Notice being given. However, in other situations, there may have been little, if any, discussion between the parties of the alleged trigger events and their consequences for the price formula prior to the Review Notice(s) being served.

In any case, even if they are not required by the parties' contract and/or do not resolve the situation, the parties' negotiations can play a valuable role in more effectively defining the claims being made and therefore the matters in dispute. Depending upon the timing of the prior steps, they may also allow further relevant data to become available for consideration by the parties before any remaining disputes between them are escalated to more formal dispute resolution.

Sophisticated and experienced parties may be tempted to conduct the early discussions and contractual negotiations themselves, without involving external experts and lawyers until later. Whilst that may save costs (if no future issues arise as a result), this period can be fertile ground for the germination of future procedural issues, particularly if, as discussions take place, further information becomes available and more detailed investigations are undertaken, the parameters of the likely disputes shift. Some development of a party's position over time is highly likely, but any inconsistencies in approach can give rise to procedural issues at a later stage.

The negotiations are sometimes to take place over a number of months and can involve meetings, presentations and exchanges of information. Consideration should therefore be given as to the status of the discussions and/or information provided to the other party. Once documents are shared with another party they will often lose any privilege that might otherwise exist. It is therefore important to decide in advance whether the discussions and exchange of information should be conducted on an 'open' basis, such that, should the parties not reach full agreement, the details can be referred to in future arbitration proceedings, or whether they should be 'without prejudice', such that they should not be referred to before any tribunal.

Questions of confidentiality and privilege must be considered in relation to the applicable law. Indeed, the significance of the parties' intentions when drafting or amending the contract, their understanding of the contract terms (as may be revealed in negotiations), their conduct after the execution of the contract or amendment (including in the negotiations), and any changes in approach as between the negotiations and any subsequent arbitration will also vary with the applicable law.

All these issues therefore need to be considered in context. Parties wanting to limit the use of information exchanged during the negotiations and avoid procedural issues relating to this arising during any arbitration will thus often be well advised to enter into confidentiality or non-disclosure agreements to define and record the position.

If the initial negotiations are concluded without resolving all the issues between the parties, there is commonly provision for further notices to be given and sometimes further discussions prior to the commencement, if still necessary, of arbitration.

2.3 Drafting of multi-tier dispute resolution clauses

Contractual provisions requiring negotiations and other steps prior to arbitration are referred to as 'multi-tier' dispute resolution clauses. Dispute resolution provisions do not always receive the close attention that they deserve, and multi-tiered provisions deserve a lot of attention. It is ironic that, in seeking to avoid disputes progressing to arbitration, providing for multi-tiered dispute resolution as a solution to that problem can often make disputes more complicated, time consuming and, therefore, costly.

An important initial question when considering including pre-arbitration steps in a contract is whether those steps should be considered to arbitration, such that arbitration can be commenced only when the prescribed pre-arbitration steps have been completed, or whether they should be obligations short of conditions. Whilst parties may wish to give pre-arbitration settlement the best chance of success, making the initial steps pre-conditions to arbitration raises the drafting and compliance stakes significantly. Failing to make clear provision for and/or to comply with such steps can prevent a tribunal from having any jurisdiction over the disputes. In some cases this can cause substantial additional costs in identifying and dealing with the failings but, in others, where it is too late to remedy the situation, the result can be that no viable claim can be pursued.

A common fault is failing to make clear what must be done or by when. For example, provision may be made for parties to seek to resolve their dispute by negotiation, but if they are unable to do so, for the dispute to be referred to the next stage of dispute resolution. Such a provision does not make clear exactly what is to be done or, importantly, over what period. This could give rise to procedural issues over both the enforceability of the required step and whether there has been compliance with the stipulation. If it is not clear whether a procedural step has been taken or complied with, then the question arises as what the consequence of that non-compliance is.

It is therefore often advisable to ensure that the pre-arbitration steps are not pre-conditions to arbitration and/or to make each step subject to a clear time limit after which the next step can be taken. Ultimately, whilst pre-arbitration steps may be useful in avoiding a protracted, formal dispute resolution process, they should not be allowed to jeopardise the final resolution of a dispute by arbitration.

3. Procedural issues arising in the arbitration

3.1 Consolidation of arbitrations

It is common for parties to have simultaneous competing claims for price reviews and the result is sometimes that each party commences arbitration against the other. Whilst it is perfectly possible to have separate arbitrations running in parallel, such a course of action could risk inconsistent decisions in light of the different submissions and evidence, particularly where the tribunals include different arbitrators. The risk can be minimised by each party nominating the same arbitrators in each arbitration or eliminated by the two arbitrations being consolidated into a single proceeding, where all the issues are dealt with together by one tribunal.

The consolidation of arbitrations is relatively straightforward where the parties are in agreement but can be more problematic where they are not. In order to assist in resolving this potential procedural issue, many arbitration rules (such as those of the LCIA and the ICC) make provision for consolidation. By adopting those rules, the parties have agreed in advance to the potential consolidation of arbitrations where appropriate (and, obviously, in accordance with the applicable rules), even if one of them has changed their mind subsequent to a dispute arising. As the disputes will normally have arisen under the same Sale and Purchase Agreement, and the same arbitration agreement will apply, there is unlikely to be an issue regarding the compatibility of the arbitrations (provided that different arbitrators have not been appointed and the consolidation is carried out at an early stage).

Where proceedings are consolidated in these circumstances, there are effectively two claimants and two respondents. For practical reasons, the parties may be designated as a claimant or a respondent but in substance each is pursuing its claim for a price adjustment and defending against the claim made by the other party. As discussed below, this may affect the way in which the procedure for the consolidated arbitration is organised.

3.2 Jurisdiction and preliminary issues

Particularly in light of the various triggers and thresholds commonly contained within price review provisions, in many gas pricing arbitrations consideration should be given to whether there are jurisdiction or other issues which should be addressed at a preliminary stage of the arbitration, prior to dealing with the remaining issues.

Jurisdiction and/or admissibility issues that relate to the pre-arbitration procedures will normally need to be raised at an early stage in the arbitration or they may be taken to have been waived. It may also make sense for them to be dealt with at that time if matters are clear or questions of contractual interpretation. For example, there may be issues as to whether any given step was complied with and/or whether it was a true pre-condition to arbitration (that is, one that prevents the tribunal from having jurisdiction), simply a contractual requirement that was breached or even a provision which is unenforceable.

There may be other issues which it makes sense to resolve at an early stage. That might be the case because they can be resolved relatively quickly and easily and, depending upon how they are decided, may affect the way (or even if) the rest of the arbitration should proceed. Alternatively, the resolution of those issues may have such an effect on the merits that resolving them is likely to lead to a settlement. For example, perhaps due to the fact that price review provisions are often heavily negotiated compromises, there are often unclear contract terms which could sometimes usefully be interpreted by the tribunal at an early stage.

Further, it is important for the parties and experts analysing a relevant market to know what that market comprises, but sometimes this can be a little vague. Is it an objective market or does it relate to the buyer? Does it cover all segments of a national market and what about exports? Similarly, what are the appropriate reference points in time for any changes to have occurred and their effects to be assessed, and the cut-off dates for the evidence to be relied upon? Gas price review arbitrations require consideration of events in the past but are generally also setting prices for the future. Where these matters have not been fully addressed in the parties' contract, they are likely to feature in arguments over which events and periods are relevant for the analysis.

Whether or not the clarification of such matters should be resolved at an early stage requires an assessment of the time and cost which it would take to resolve them and the savings in time and cost which might flow from that resolution, as against the time and cost involved in proceeding without the clarification (when, for example, alternative approaches may be necessary or the parties and their expert might fail to address the same points and pass like ships in the night).

Other issues relating to the power of the tribunal to adjust the price formula may be so wrapped up with the merits of the dispute that it does not make sense to resolve them separately, in advance of other substantive issues. For example, if the adjustment (or not) of the price formula depends upon the effect of disputed market changes, the complexity of those issues, and the need for factual and expert evidence will often mean that they cannot sensibly be hived off for early determination.

3.3 Organisation of written submissions and evidence

Apart from the question of whether to deal with any jurisdiction or other issues at a preliminary stage, the parties and/or tribunal will need to consider how best to organise the remaining procedure for the arbitration. This can be done in many ways and should normally be discussed soon after the constitution of the tribunal. Some arbitration rules, such as those of the London Court of International Arbitration (LCIA), make provision for service of written submissions following the appointment of the tribunal absent agreement or a direction from the tribunal to the contrary. The remainder of the procedure remains to be fixed in parallel or with the benefit of sight of those submissions.

Other arbitration rules, such as those of the ICC, leave it to the parties and tribunal to sort out the entire procedure and require that a case management conference is convened at an early stage for that purpose. Under most international rules, however, the parties and tribunal are reasonably free to decide upon the desired structure and timing for the arbitration.

Where, as discussed above, both parties bring claims for a price review (eg, one seeking an increase and the other a decrease in the price of gas) and therefore consider that they are, in substance, both a claimant and a respondent in the arbitration, this may raise the issue of whether there should be simultaneous rather than consecutive submissions. Simultaneous submissions mean that each party will be presenting its claims and then responding to those of the other party simultaneously, which can save time over the more standard, consecutive, procedure where the claimant proceeds first, followed by the respondent.

Another basic procedural question is whether to have written submissions (generally together with documents relied on) served in advance of, or together with, factual and expert witness statements (the latter approach often being referred to as 'memorials'). Given the often heavy reliance on expert evidence in price reviews, in many cases the memorials approach may be preferred. On this basis, each party will prepare and submit all its submissions and evidence (documentary, witness statements and expert reports) together. This requires more time for preparation but allows the claims and evidence to be prepared and viewed at the same time.

Given the need for each party to respond to the submissions and evidence prepared by the other party, there are usually at least two rounds of memorials, and sometimes three or even four. If there is to be a disclosure stage, this is commonly scheduled after the first round of memorials has been submitted by each party, so that the requests for documents can take account of the submissions made (in the same way that disclosure usually follows the written submissions when those are split from the service of witness statements and expert reports).

3.4 Disclosure and data

There is usually only limited factual evidence (contemporaneous documents and witness evidence) in gas price review arbitrations, with the emphasis being largely upon expert evidence. However, provision is typically made for a limited disclosure exercise in case that is appropriate, for example by way of requests (supported by evidence/arguments) for specific documents or categories of documents that are relevant to the issues in dispute.

Often more important than witness evidence and contemporaneous documents is the evidence relied upon by the experts, some of which may be publicly available and some obtained from private sources. Of course, particularly where an expert relies upon data that is not publicly available, it is important that this is made available to the other participants in the arbitration. However, this can give rise to issues, especially where it is not clear precisely what data has been relied on, and whether this is selective rather than a full data set.

The private and confidential nature of the data used in price review arbitrations can also give rise to procedural challenges. There are data protection issues related to information about end users and their contracts, and there can also be considerable commercial sensitivities, particularly where the price review concerns subjective elements, such as the profitability of the contract for the buyer. This sensitivity is even more acute when the parties to the price reviews are in competition with each other in the relevant end-user market, which is often the case.

The sensitivities surrounding data often results in parties and/or the expert they have engaged being given access to data in protected data rooms rather than being given possession of the data. This can give rise to issues over any restrictions on the access granted, for example, as to time and/or how the data can be reviewed. Many of these practical issues are best resolved between the parties and their advisers as far as possible, but recourse to the tribunal is, inevitably, sometimes required.

3.5 Dealing with expert evidence

As price review arbitrations are usually heavily focused on expert evidence, there is often significant discussion over the way in which the evidence should be dealt with. Parties each normally engage independent experts and it is rare for the tribunal to appoint its own expert. This places a considerable burden on the tribunal to get to grips with the complex subject matter at issue, and the experience and ability of the arbitrators to do this should clearly be an important factor in their selection.

In order to try to ease the burden on the tribunal, the experts are sometimes asked to liaise so as to identify between them the areas of agreement and dispute respectively, sometimes producing a joint statement to this effect for the tribunal. This can give rise to issues as to the extent to which the parties and their lawyers should be involved in joint meetings and the joint statement production process, but this is normally worked out with the experts without the need to involve the tribunal.

Another approach which can be helpful is to allow each expert a period at the start of their giving evidence at the Hearing to make a presentation on the main points on which they believe the tribunal should focus. The main procedural concern in this context is to ensure that the experts do not introduce any new data or analysis at such a late stage without prior approval having been obtained. If hard copies of the materials supporting the presentations are exchanged before being delivered, that issue can be avoided (or at least the potential ambush can be considered in advance and the impact minimised).

Expert witness conferencing, sometimes also called 'hot tubbing', is a process often used in gas price arbitrations. This can take many forms but the common feature is that the experts from both sides give at least some of their evidence together. The tribunal may take the lead in asking the experts questions, with the parties being given the opportunity subsequently to ask any remaining questions. However, on other occasions, the parties may go first, and the tribunal asks any questions it wishes as the cross-examination proceeds or afterwards.

Although not specific to gas pricing arbitrations, in recent years it has become more common for expert reports to be drafted by two experts rather than one. In such cases, procedural issues can arise as to how their evidence is to be dealt with at the hearing. Commonly, the experts who wrote the report are questioned together. Where they have separate responsibility for different parts of their report, any given question may be directed at the relevant expert. Sometimes, however, the experts may have areas of overlap or have produced the whole of their evidence jointly. In such event, rules may be agreed or imposed as to how questions may be addressed or answers given.

3.6 The scope of the award

An unusual feature of gas price review arbitrations is that the dispute is over what, if any, adjustment to the parties' contract should be made rather than whether the contract has been breached. The tribunal is asked to decide whether an adjustment to the price formula should be made and, if so, to make it.

Issues can arise as to the scope of the tribunal's power, both in relation to what the parties have each claimed should be done and as regards whether the price review provisions permit the degree of alteration contemplated.

The making by the tribunal of an award which differs from the relief sought by either party can give rise to issues in any arbitration. However, this may be considered a particular issue in circumstances where the tribunal is fixing the future business relationship between the parties by amending the pricing formula in their contract, which is a fair way away from the normal domain of contentious lawyers. The issue may become even more acute where the tribunal is considering making a substantial change to the price formula, rather than a simple adjustment. Ultimately, this is a question of the tribunal's power as derived from the parties' contract and applicable law.

4. Concluding remarks

Gas price review arbitrations form a fascinating sub-group of arbitration that gives rise to a host of procedural and substantive issues. The important nature of the supply contracts behind these disputes and the huge sums often at issue justify significant focus on the arbitration process and the best way to minimise and resolve the procedural issues that arise.

The complex and multi-tiered nature of many price review disputes, and the very nature of contentious lawyers, means that some procedural challenges are perhaps inevitable. However, these issues can be minimised, and certainly the outcomes made more predictable, by careful drafting at the contracting stage, informed by an appreciation of where the problems can arise.

Evidence in gas pricing arbitration

Augustin Barrier
Noradèle Radjai
LALIVE

1. Introduction

Counsel and tribunals in gas pricing disputes are confronted with the tension inherent in long-term gas supply agreements ('GSAs') between the need for predictability, which mandates high thresholds for seeking and establishing the need for a price revision, and the concurrent need for flexibility, which requires that GSAs may be adapted to take into account changes in energy markets that have all too often gone through phases of turmoil. This tension is compounded by the limited availability of data regarding market prices and the presumption that, the parties being sophisticated and prudent operators, they are deemed to be aware of the prevailing market conditions.

This chapter examines the specificity of evidence in gas pricing arbitration and in particular how it may depart from principles commonly applied in other commercial disputes.

Indeed, although the burden and standard of proof are determined pursuant to applicable law, their regime may vary as a result of the specific language of the GSA at issue, and are in addition impacted by presumptions inherent to the industry.

The scarcity of data available in gas pricing disputes also impacts the selection by the parties of available evidence in support of their respective positions.

Finally, the focus of gas price reviews on a determined temporal scope set usually by the reference period raises issues as to the admissibility of evidence from outside this determined scope. This touches on the admissibility of evidence post-dating the end of the reference period, the evidence from the negotiation phase, and also arbitral decisions rendered in previous matters.

2. Burden and standard of proof in gas pricing arbitration

The determination of the party bearing the burden of proof, as well as the standard of proof, in gas pricing disputes, follows from the applicable law, as in any commercial dispute. However, certain specificities arise in gas pricing arbitrations, in which the requesting party has to meet a burden articulated in two steps, one at the request stage and the other at the adjustment stage.

Furthermore, both the burden and standard of proof are affected by certain presumptions specific to gas pricing disputes.

2.1 Burden of proof

The prevailing principle in international arbitration is the principle *actori incumbit probatio*, that is, the party making an allegation must prove it.[1]

In the context of a gas price review arbitration, the burden of proof rests on the party seeking a revision,[2] as in any dispute. However, this matter is rendered somewhat more complex in gas pricing disputes since the requesting party must in fact meet two separate burdens. As a first step, it must prove in its price revision request that the contractual requirements for a price revision are met. As a second step, it must prove, in the negotiations or, should those fail, in the subsequent arbitration, that a revision of the contract price is warranted.

Unlike the burden of proof of a party's claim for an adjustment of the price formula, the burden of proof at the stage of a price revision request is often more formalistic and may be more or less stringent depending on the GSA. However, this stage is not simply procedural. Normally, the requesting party must show that a so-called trigger event has occurred and explain the consequences that this has on the pricing terms in force. The burden of proof in this phase is usually lower than in the adjustment phase, requiring a *prima facie* demonstration of the merits of the price review requested. This step can nonetheless be crucial in gas pricing arbitrations as the issue of whether a price review request has been served in accordance with the GSA and adequately substantiated may have an impact on the admissibility of the requesting party's substantive claims in the arbitration. Irrespective of the applicable law, this principle derives from the GSA, which usually provides that the party requesting a revision must substantiate its request.

The language for this requirement varies from one contract to another. The threshold is generally limited to providing the reasons for the revision (eg, by stating that the requesting party should "inform the other party in writing, stating the reasons therefore and proof of these reasons" or "notify the other party in writing of a price revision request, substantiating the grounds for its request").

Other GSAs even provide that the trigger letter should be accompanied by supporting evidence. This was the case in ICC Case No 9812, in which the contract specified that: "[e]ach party shall provide all necessary information to substantiate its own claim. No Party shall be required to disclose any business

---

1 J Waincymer, *Procedure and Evidence in International Arbitration* (Wolters Kluwer, The Netherlands, 2012) 763; N Blackaby, M Hunter, C Partasides and A Redfern, *Redfern and Hunter on International Arbitration* (OUP, United Kingdom, 2015) para 6.84.

2 See Final Award in Case 9812, August 1999, "An application of Article 6.10(a)(1) implies that the burden of proof is on the party requesting a price review" (2009) 202 *ICC International Court of Arbitration Bulletin* 81.

secrets nor [sic] to provide such information as the other party may need to substantiate its claim."[3]

Some GSAs merely require that the requesting party notify the other party of its price revision request. Yet, even then, the GSAs often require that the requesting party substantiate its position in the ensuing negotiation period. Such substantiation may include a description of the changes invoked, with reference to economic analyses or data showing the need for a revision.

Once the request has been admitted, either voluntarily by the receiving party or through arbitration, the requesting party must establish that the price revision it is seeking is justified, for instance because the contract price has become out of sync with the parties' original bargain. In this phase, the requesting party has the burden of proving that, during the contractual reference period, the economic circumstances changed significantly compared to those that prevailed at the time the parties last agreed on the pricing terms of their GSA, and that these changes are not reflected in the pricing provisions in force so that an adjustment is required.

The exact scope of the requesting party's burden of proof with respect to the adjustment phase is usually defined by the price revision clause. In particular, the clause will determine:

- the reference period to be taken into account (usually the period of time between the last price revision and the request for a new price review);
- the reference market, which may be defined in geographic terms (eg, the market of the buyer or a wider market), sectorial terms (eg, the electricity, industrial or retail markets) or a combination of both;
- the specific circumstances that must be examined, if they are specified at all (eg, gas prices and their structure, the price of alternative energies);
- the magnitude of the required change in economic circumstances (eg, whether the change must be "substantial", "material" "unforeseeable", "outside the parties' control");
- the impact of the change on existing pricing terms; and
- whether new or alternative pricing terms reflecting the changes may be introduced.

Although the burden of proof on whether a price revision is warranted cannot shift, the more the requesting party substantiates its request, the more the party resisting the price revision will have to present contrary evidence to counter it.[4] Similarly, the party resisting the price revision may present its own case in response to the claimant's claim (eg, by proposing a different

3 *ibid*, 79.
4 J Waincymer, *Procedure and Evidence in International Arbitration* (Wolters Kluwer, The Netherlands, 2012) 763.

interpretation of the reference market or alternative benchmarks to be taken into account to track the prices of natural gas, LNG or alternative fuels in such markets). In such cases, the respondent will bear the burden of proving this positive case.

2.2 Standard of proof

Tribunals often determine the degree of evidence required in accordance with the nature of the proposition to be proved. As a general rule, "the more startling the proposition that a party seeks to prove, the more rigorous the arbitral tribunal will be in requiring that proposition to be fully established".[5]

The standard of proof applied by tribunals in gas pricing disputes is also contingent on the availability of the evidence. More often than not, proving the need for a price revision requires reliance on pricing data, which is either not publicly available or incomplete or confidential. It has been suggested, that, in such cases, the standard of proof should be adapted accordingly, that is, that the tribunal should accept that a party has met its burden of proof if it has adduced sufficient evidence in light of what is available to it.[6]

2.3 Presumptions in gas pricing disputes

Certain presumptions arise in the context of evidence in gas pricing disputes, not by virtue of legal requirements, but as a result of specific pricing terms or industry practice.

A key presumption is that the parties are prudent and sophisticated operators. This presumption is logical in light of the technical and legal skills required to be able to even enter into a GSA.

This presumption is sometimes formalised in GSAs. For instance, in ICC Cases Nos 13504 and 9812, the contract provided that the parties' conduct should be assessed under the standard of a "prudent and efficient gas company".[7] Other contractual terminologies include "reasonable and prudent operator", "efficient and prudent operator" or "skilled and efficient operator". Some GSAs even provide a definition of the standard, often by making it clear that the parties are deemed to perform their obligations in good faith, exercising a degree of skill, prudence and foresight to be expected from an operator engaged in the same type of undertaking.

Under this presumption, the tribunal will consider both parties to be professionals, familiar with the natural gas market and industry.

---

5 N Blackaby, M Hunter, C Partasides and A Redfern, *Redfern and Hunter on International Arbitration* (OUP, United Kingdom, 2015) paras 6.86–6.87.

6 See P Pinsolle, "Confidentiality in Gas Price Reviews" in M Levy, *Gas Price Arbitrations* (Globe Law and Business, London, 2014) 57–59.

7 Final Award in Case 13504, January 2007, (2009) 202 *ICC International Court of Arbitration Bulletin* 105–06; Final Award in Case 9812, August 1999, (2009) 202 *ICC International Court of Arbitration Bulletin* 79.

This may have practical consequences in terms of the conduct and evidence expected from the parties. For instance, in ICC Case No 13504, the tribunal considered that this standard could influence the appreciation of whether a change in economic circumstances was outside the parties' control.[8] This principle also guided the tribunal's analysis of the method and evidence proposed by the claimant to calculate the value of natural gas in the relevant market.[9]

A second presumption, which derives from the first one, is that the parties, as prudent operators, were fully aware of all the prevailing economic circumstances when they agreed on the terms of the GSA or renewed their consent to the pricing provisions.

The practical consequence of this presumption, in terms of evidentiary standards, is that any contemporaneous publicly available evidence could be presumed to have been known to, and taken into account by, both parties. Consequently, a party seeking to argue that it was not aware of a certain event, which was discoverable from the public domain will have to meet a very high burden of proof.

A third presumption, which derives from the first two, is that, when entering into a GSA or renewing their agreement on its pricing provisions, the parties will not only have made the necessary inquiries regarding the prevailing economic circumstances, but also made reasonable projections as to how these circumstances would subsequently evolve.

In light of the volatility of the oil market and the changes that have affected the gas markets over the last two decades, this presumption is obviously not conclusive. However, certain provisions, like hardship clauses or the law applicable to some GSAs,[10] enable a party to seek a price revision should an unforeseeable change in circumstances arise, outside the parties' control. In such situations, if the data available the last time the price was agreed would have enabled the parties to reasonably anticipate the change invoked by the requesting party, tribunals may reject the price revision request.

Naturally, all these presumptions are rebuttable, which in turn implies that the burden of disproving them would lie on the party seeking to do so.

8 Final Award in Case 13504, January 2007, (2009) 202 *ICC International Court of Arbitration Bulletin* 108 (at para 110).

9 *ibid*, pp114–115 (at para 138). In particular, the tribunal considered whether the claimants had acted as a "prudent and efficient gas company" in order to determine whether the prices presented by the claimants as obtainable in its home market were indeed market-reflective or the consequence of events within its control, ie significant discounts off market price to its customers, as purported by the respondent. In so doing, the tribunal noted that "Claimants have been most reluctant to disclose information relating to the relations between Claimants and their traditional customers".

10 See M Polkinghorne, "Changes of Circumstances as a Price Modifier" in M Levy, *Gas Price Arbitrations* (Globe Law and Business, London, 2014) 63.

3. Types of evidence used in gas pricing disputes

A specific trait of gas pricing arbitrations is that they typically do not arise from a breach of the GSA, but from a dispute over whether a specific set of circumstances justifying a price revision has occurred and whether a price revision is warranted under those circumstances. In other words, the economic circumstances can play a much greater role than the factual aspects of the parties' relationship. As a result, although parties to a gas pricing arbitration will rely on documentary evidence, experts and fact witnesses, as in other international disputes, the relative focus and importance of such evidence is rather particular in these types of disputes.

3.1 Documentary evidence

The key issue in relation to documentary evidence in gas pricing disputes is the availability of reliable data concerning the value of natural gas, or competing fuels, in a specific market. This data, although necessary to determine whether a change in economic circumstances requiring a price revision has occurred, remains scarce.[11]

Whether introducing this evidence through an expert or directly,[12] parties often use combinations of different sources of publicly available data (statements by energy regulators, statistics, market reports, reports from other gas companies to their shareholders, etc) to get as close as possible to the reality of the value of gas in the market. Some practitioners have compiled impressive lists of available public sources that can be used to that effect.[13] However, the calculation methods may be unknown or inconsistent from one source to the next. The use of public data is often challenged in arbitrations on this basis. The most relevant data are sometimes the prices of gas under individual contracts, which remain confidential.

Public statements by gas companies that they are requesting price revisions under their GSAs can also constitute possible indicators of change in the market even though the provisions of the underlying contracts, and, in particular, the conditions for price revisions remain confidential.[14] Publicised results of gas pricing arbitrations provide similar indicators. Multiple statements by buyers that they have succeeded in securing a price revision may assist other buyers in

---

11 There is undeniably a trend towards more transparency in the natural gas market. The availability of market-based hub prices provides useful indications concerning the value of gas. The time may come when gas market hub prices will be applied as the relevant benchmark. Be that as it may, for the time being, the market is not changing uniformly. Although Nortwest European hubs are increasingly mature and market-reflective, the same cannot be said for the rest of the European market and, to a larger extent, the Southeast Asian market.

12 Parties often involve consultancy firms as expert witnesses to provide in-depth analysis of the value of natural gas in a given market based on independently gathered data. Expert evidence is discussed further in section 3.2 below.

13 J Bowman and J Mattamouros, "Resources on Resources: Publications and Databases Available to Help Advocates in International Oil & Gas Disputes" in JM Gaitis (ed), *The Leading Practitioners' Guide to International Oil & Gas Arbitration* (Juris, 2015) 937 *et seq.*

14 M Polkinghorne and S-M Volkmer, "Price Re-Openers in Long-Term Gas Supply Agreements" in M Gaitis (ed), *The Leading Practitioners' Guide to International Oil & Gas Arbitration* (Juris, 2015) 550.

convincing tribunals that their own claim is justified. Similarly, statements providing information as to changes in formula indexation often serve as an argument in subsequent gas pricing disputes to request a switch from oil to hub indexation.[15] However, any evidence of this nature must be approached carefully in view of the diversity across GSA provisions regarding price reviews and the basis on which the price may be adjusted.

Another issue often addressed by arbitral tribunals in gas pricing disputes is the asymmetry in access to evidence. Buyers often enjoy access to the data provided from their marketing department in relation to obtainable prices in the market (records of realised and lost sales). Sellers, unless they have downstream capacities or subsidiaries operating in the relevant market, are less able to provide their own price analysis. Nevertheless, if they have multiple customers in the market, they have knowledge of the prices charged to each of them. From both the buyer's and the seller's perspective, this data is often confidential and may not be presented to the tribunal in the form of direct documentary evidence (subject to a document production order, as discussed below).

The same question arises in relation to the parties' profits. The respective profits made by either or both parties may be relevant depending on the price revision provision. Such 'in any case' clauses provide that either the buyer or both parties should be able to market the gas at a reasonable profit. Likewise, the setting of a floor in the pricing provision may be indicative of the parties' intention to ensure that the seller makes sufficient margins or recoups its costs. Some contracts also provide that the outcome of the price revision must be 'fair' or 'fair and equitable', which may require the requesting party to show that its proposed revision enables both parties to earn a reasonable profit from the contract. Because they involve transactions with third parties, such profit figures are also often confidential.[16]

The document production phase, when not prohibited under the GSA,[17] may provide an avenue for establishing more even access to evidence by obliging the parties to produce or make available for inspection data that would otherwise not have been available on record. The sensitivity of such data often results in acrimonious disputes between the parties over document production, requiring the tribunal to intervene in order to ensure that the said data's confidentiality is preserved.[18]

15 See eg the press release by RWE on 1 July 2013 informing the press that an arbitral tribunal had ordered a significant price reduction and introduced a new gas market indexation, which was widely invoked in subsequent arbitration proceedings.

16 However, certain consultancy or audit firms may review and provide an anonymised account of such confidential data. Such expert evidence is discussed further in section 3.2 below.

17 Some gas supply agreements specifically limit the possibility of parties to request any kind of documents from each other, or provide that the parties should not be required to produce confidential information. However, these clauses, which often relate to the request or negotiation stages, are not necessarily implemented by arbitral tribunals in the course of the arbitration.

18 P Pinsolle, "Confidentiality in Gas Price Reviews" in M Levy, *Gas Price Arbitrations* (Globe Law and Business, London, 2014) 57.

Documentary evidence may also feature in gas pricing arbitrations in relation to the interpretation of the GSA or the assessment of the parties' conduct. Documents pertaining to the negotiation of the terms of the GSA (minutes of meetings, presentations, correspondence) may shed light on the intention of the parties in relation to the meaning of certain provisions, if they are admissible under the applicable law. Likewise, records of the positions taken by the parties in previous price review negotiations may also assist the tribunal in assessing the parties' intentions as to the meaning of certain terms of the GSA.

3.2 Experts

Most commercial disputes involve the submission of expert evidence. This evidence is often limited to discrete technical issues or, more generally, the quantum of the claims and counter-claims.

By contrast, the role of experts in gas pricing disputes is central to the case as a whole. This results from the importance of economic analysis and industry practice in the operation of long-term GSAs. As noted above, the difficulty of procuring or presenting direct documentary evidence in arbitration renders the role of expert evidence crucial.

In the context of the arbitration itself, the term 'experts' may cover many different aspects. It is common for the parties to rely on one expert from one consultancy firm to cater for all aspects of the case. An alternative that has developed in recent years has been to resort to several different types of expert covering different aspects of the dispute. For instance, the expert teams may comprise:

- economists specialising in the valuation of natural gas or LNG, dealing with the analysis of the market and possible changes affecting it, as well as the proposed price adjustment;
- economists specialising in competition issues, given the increasing weight given to competition law issues in long-term GSAs, and especially in LNG contracts;
- former executives from the industry who have become independent consultants and may testify to the dynamics of a given market, industry practice, or their own experience in relation to price reviews based on similar provisions;
- accounting or audit firms in charge of providing analysis of the parties' internal accounting data, especially in cases where the parties' ability to

19 Some cases involve even more specialised experts. For instance, a party may require the services of experts in charge of quantifying changes in the price of gas on the economy of an entire country. This is achieved by building a mathematical model reproducing all sectors of a country's economy with all their variables so as to measure the quantitative effect of modifying one of these variables, the price of natural gas. The size of this model and the quantity of data required obviously limits the number of experts available to perform such valuations.

make profits is disputed, or presenting in an audited form information that would otherwise not be produced because of its confidential nature; or

- consultancy firms offering quantitative analysis of market data on gas prices, including by providing in an anonymised form data gathered independently in the market pertaining to confidential transactions.[19]

Submitting evidence from several experts whose scope of expertise may largely overlap inevitably requires a higher degree of coordination between them, and with the parties and their counsel. On balance, the risks of overlap and minor contradictions between experts in different fields may however be less problematic for a party than the risk of the tribunal perceiving its only expert as lacking credibility in a specific area of importance to the dispute.

It is legitimate for a party to use an expert's knowledge to its full extent. For instance, parties may be tempted to ask their expert to opine on the interpretation of a price review clause. This is usually not inappropriate: industry experts may provide useful insight on the meaning of technical terms, such as 'market value of gas', 'obtainable prices', or 'gas-to-gas competition'. However, experts should in most cases not be asked to opine on the operation of the clause, notably in relation to whether a price revision is warranted in light of its terms.

3.3 Fact witnesses

Given the prevalence of the economic background in adjudicating price review disputes, the factual background of the parties' relationship is usually of limited relevance. Evidence from fact witnesses may nonetheless prove necessary in some situations.

First, witnesses may shed light on the negotiating history of the contract and provide insight as to what the parties meant by certain terms and the articulation of certain clauses, as well as the contractual bargain struck by the parties at the outset. Whether this approach is admissible ultimately depends on whether, and the extent to which, the applicable law admits such evidence for interpretation purposes. In the context of a commercial arbitration, arbitrators may apply the issue of admissibility with some flexibility, using the applicable law rather as guidance as to the weight to be given to such evidence.

Secondly, individuals who took part in the negotiations following the price revision request may testify to the conduct of these negotiations. This is of course subject to the negotiation period not being covered by a confidentiality arrangement.[20] In particular, a witness may speak to the basis of the price revision request and its initial scope (in case the scope varied in the arbitration),

20 As to the admissibility of evidence from the price review negotiation phase, please see section 4.2 below.

or whether the parties were in agreement as to the relevance of certain data or markets to the revision of the price.

Thirdly, the buyer may present witnesses to testify to the real-life conditions in the market, notably on the relationship with customers in various market segments. For instance, a witness from the buyer's marketing department may describe discussions and negotiations with its customers to show that the existing contract price no longer enables the buyer to market the gas economically. In response, the seller may seek the testimony of local private gas marketers or, if it has downstream capacity, members of the marketing department of one of its subsidiaries.

Fourthly, the seller may ask some of its employees in charge of managing its portfolio in a given market to provide evidence on the relationship the seller has with other market players. The seller may seek to prove that the other buyers in the market have been satisfied with negotiated solutions, be it through a price reduction or additional volume flexibility through temporary reductions of their take-or-pay obligations. If the buyer requesting the price revision has refused similar proposals, this may help depict its attempt to seek relief through arbitration as rigid and unreasonable.

4. Admissibility of evidence

Because the admissibility of evidence in gas pricing arbitration is often confined to a specific period or contract, the issue arises whether evidence from beyond that scope is admissible. Typically, gas price review arbitrations may give rise to such considerations regarding the temporal scope of evidence, and whether or not evidence from the negotiation phase or previous arbitrations can be introduced.

4.1 Temporal scope of evidence

A common feature of gas price review arbitrations is to determine whether a change occurred in the market that would justify a revision of the contract price. To establish this, the claimant must usually identify such a change during the reference period. The starting date of the reference period is usually the effective date of the last price revision, or the date on which the GSA was entered into. The end date of the reference period is typically the effective date of the new price revision. This latter date should in principle also determine the cut-off date for data submitted by the parties concerning the alleged change in the market, since data after this date may no longer be relevant to establish a change during the reference period.

This does not mean, however, that such later evidence would automatically be rejected.[21] Arbitral tribunals have often been reluctant to consider the

21 See P Pinsolle, "Confidentiality in Gas Price Reviews" in M Levy, *Gas Price Arbitrations* (Globe Law and Business, London, 2014) 60–61.

effective date as a complete limitation on the admissibility of evidence. The assessment of evidence post-dating the price revision request must be evaluated on a case-by-case basis. For instance, data which was published only after the end of the reference period may be relevant to determine changes that occurred during the reference period. It may also be arguable that the GSA allows the tribunal to consider the outcome of the revision, for example, by providing that the result of the revision must be fair and equitable. More practically, the tribunal may simply want to conduct a 'sense check' as to whether its proposed price revision is sensible when viewed against price data post-dating the effective date. Such considerations can provide a basis for the admission of evidence post-dating the reference period.

A claimant may also seek to adduce evidence relating to the reference period if that was published or became available only in the course of the negotiations or arbitration. This is not uncommon especially where a party is required by the contract to make a price revision request by a certain date. In such a case, when the date of the request coincides with the end of the reference period, it will not have had access to all of the information for the reference period relevant to its claim.

Although the admissibility of such additional evidence is usually uncontroversial, since it still concerns data for the reference period, it may raise concerns if the new information requires the claimant to amend its claims or make new claims. The respondent may object to the admissibility of such new claims on the ground that the scope of the dispute is restricted by the terms of the initial request.

Moreover, the admissibility of such new evidence will likely depend on the stage of advancement of the proceedings. In particular, before admitting such evidence, the tribunal will consider whether the other party will have sufficient time to review and provide its comments on the data submitted. In case of a belated submission that prejudices the process, the tribunal will in all likelihood not admit the evidence or, if it does, it might discount the weight to be given to such evidence in reaching its decision.

4.2 Evidence from the price review negotiation phase

The admissibility of evidence from the negotiation phase raises particular issues.

Whether these documents, and the negotiation process in general, are inherently privileged is a hotly debated issue. The answer varies significantly depending on the legal system of reference and the circumstances of the discussions (eg, whether lawyers, external or in-house, were involved).

Over the course of negotiations, parties often adopt positions that are less developed and – at least in principle – more constructive than during the arbitration where the issues in dispute are more crystallised. For instance, the seller may acknowledge the existence of a change, albeit transient, in the market and propose concessions such as a downward adjustment to the take-or-pay requirements or a slight reduction of the price through a reduction of the base price in the formula (P0). The buyer may make similar concessions.

In the absence of appropriate confidentiality provisions, it is likely that evidence of such concessions in the negotiation process will surface in the subsequent arbitration. Accordingly, if a party wishes to avoid this, it is often preferable to enter the negotiation with a non-disclosure agreement in place.[22]

Another issue that may arise in connection with evidence from the negotiation phase is that a party may dispute the admissibility of a claim in the arbitration because the requirement that the parties negotiate in good faith has not been met. In this scenario, the parties may have to waive any confidentiality that may apply, to produce materials from the negotiations, in order to prove whether or not the negotiation was pursued in good faith.

Even where evidence from the negotiations is admitted, for instance where there is a waiver of confidentiality, or where the negotiations are not protected by confidentiality or privilege, the question still arises regarding the weight to be given to evidence adduced from the negotiations. The negotiation positions taken by the parties in a spirit of cooperation should in principle not preclude them from advancing different arguments in the subsequent arbitration.[23] Nonetheless, tribunals may draw negative inferences from a party's significant shift in position between the negotiation and the arbitration.

4.3 Previous arbitral decisions

Arbitral awards rendered in gas price reviews have proliferated over the last decade or so, but almost all of them remain confidential. This raises the issue of the admissibility and effects of previous arbitral decisions in gas pricing disputes.

(a) Previous decisions between the same parties

In long-term contracts, it is not unusual that the parties go through several price revisions that result in arbitration proceedings and arbitral awards. The admissibility of these awards is not controversial insofar as their production does not raise confidentiality concerns.

The problem rather rests with the impact these previous decisions have on the subsequent tribunal's ability to decide the case and their preclusive effects on the parties as a *res judicata*. Since issues of preclusion are generally considered to be procedural rather than substantive in nature, the law of the seat will tend to govern the effect of the previous award,[24] although this will depend on the applicable substantive law and the law of the seat.

---

22 *ibid*, at pp48–50.

23 See N Radjai and J Landbrecht, "Arbitration and price review negotiations: previous arguments no longer relevant?" (ILO, 17 August 2015).

24 See International Law Association, "Interim Report: '*Res Judicata* and Arbitration'" (2004) 26. See also C Seraglini, "Le droit applicable à l'autorité de la chose jugée dans l'arbitrage" (2016) 1 *Revue de L'Arbitrage* 51.

In most civil law systems like France, Belgium, Switzerland, Germany or Sweden, *res judicata* only applies to the actual dispositive findings of the tribunal, not the reasoning behind these findings.[25] In common law systems, *res judicata* may prompt preclusive pleas such as cause of action estoppel, issue estoppel and abuse of process, which may apply to any issue decided by the first tribunal which is necessary for deciding the claim in question.[26] There may thus be differing approaches to this question depending on the applicable law.

The uncertainty of the effect of an earlier award on a later tribunal is compounded by the fact that price reviews, even under the same contract, arise in different time periods and in different market circumstances. Given the typical length of intervals between price revisions in long-term contracts, successive arbitration proceedings are rarely initiated in the same context. For instance, the factual and economic background of an award rendered in relation to a dispute pertaining to the increase in oil prices in 2009 is very remote from that of a dispute relating to the decoupling between oil prices and hub prices in 2011–12.[27] Can the interpretation of key contractual provisions be separated from the context in which they were agreed? And if so, will the conclusions derived by one tribunal from such an interpretation remain valid by the time the next tribunal has to decide on different claims based on the same provisions? In other words, the requesting party cannot necessarily consider that its burden of proof is satisfied because certain issues have been addressed by a previous arbitral award in a different context.

This is not to say that the second tribunal should disregard the first award altogether. Whichever law is applicable, the operative part of the award should bind the next tribunal. Furthermore, the findings of the first tribunal may be taken as a starting point by the subsequent tribunal. However, given the fast-paced evolution of international gas markets in the last two decades, then subject to the applicable law, subsequent arbitral tribunals should remain at liberty to depart from findings of fact and law, especially if they are no longer adapted to the economic context.

In any event, regardless of the position a tribunal reaches on the *res judicata*

---

25 B Hanotiau, *Complex Arbitrations: Multiparty, Multicontract, Multi-Issue and Class Actions* (Kluwer, The Hague, 2006) para 534. See also K Hobér, *International Commercial Arbitration in Sweden* (OUP, United Kingdom, 2011) para 7.117; L Radicati di Brozolo, "*Res Judicata* and international arbitral awards" (2011) 38 *ASA Special Series*.

26 B Hanotiau, *Complex Arbitrations: Multiparty, Multicontract, Multi-Issue and Class Actions*, (Kluwer, The Hague, 2006) paras 523–29. See also *Arnold v National Westminster Bank plc* [1991] 2 AC 93 at 105E.

27 For instance, one of the very few public awards, rendered in ICC Case No 9812, was rendered in 1999 (see Final Award in Case 9812, August 1999, (2009) 202 *ICC International Court of Arbitration Bulletin*). The term 'value of gas' in the contract was interpreted by the tribunal as meaning the price which "makes an end-used indifferent with respect to the choice of gas compared with the choice of the most competitive alternative energy available to the end-user" (*ibid*, p75). This economic concept, known as the 'indifference principle' remains relevant for insulated markets. However, the same contractual terms would no longer be considered as referring to this concept in a Western European context. In fact, in a subsequent arbitration based on the very same contract, in ICC Case No 13504, the tribunal upheld a different interpretation of the same contract terms.

effect of the previous award, it may be mindful of the position taken by the parties in the arbitration in relation to issues that remain unaffected by the economic context.

Parties would therefore be well advised to consider whether the positions taken in a given arbitration could not later be used against them in a subsequent arbitration, in which the context may be different and roles reversed.

(b) Previous decisions with different parties on similar GSAs

Although the record of publicly available anonymised commercial arbitration awards is growing,[28] only very few awards rendered in gas pricing disputes have been made public, most likely because of the sensitive information these contain, including the outcome itself which a party may have an interest in shielding from its customers, suppliers and competitors.

However, since parties to price revision disputes are often 'repeat players', with repeat experts, and even repeat arbitrators, these groups may over the years end up with a unique private record of arbitration awards.

Parties may thus wish to use previous awards as evidence in subsequent proceedings. Sellers who seek to harmonise the contractual provisions throughout their portfolio may have an interest in ensuring that these provisions are interpreted uniformly by producing excerpts of previous awards. Likewise, buyers who have obtained favourable decisions against sellers based on changed conditions in their market will consider these awards as relevant to subsequent arbitrations. However, the issue of the confidentiality of these materials will arise. Indeed, if an award is presented unredacted, it is likely to be considered as infringing the rights of the third party involved. The party seeking to produce the award may seek a waiver from this third party, but is doubtful whether it may secure it. Conversely, in case the award is produced redacted, the other party may claim that it was deprived from the benefit of reviewing it in context so that due process may be affected in case inferences are drawn therefrom.

In the same vein, a party may request that the other party produce awards rendered in or materials from other cases, if it considers that it may help its case. Here again, it is likely that the party requested to provide such sensitive materials will decline from doing so on the ground that they are confidential.

Either way, the tribunal will thus have to decide whether to admit such evidence based on the applicable law and procedural rules.

A review of the practice adopted in several gas pricing disputes shows that tribunals are usually reluctant to order a party to produce previous unrelated

28 For instance, the *Journal du Droit International*, the *ASA Bulletin* or the *ICC Bulletin* publish excerpts of arbitral awards on a regular basis. Likewise, a large number of investment arbitration awards are publicly available on the ICSID website or other specialised publications.

awards, including excerpts or redacted versions. Beyond confidentiality, tribunals may have concerns regarding the relevance of other awards under different GSAs with different price review provisions and potentially relating to different markets. Tribunals may also be wary that such awards may be used by the party requesting their production for purposes other than the arbitration, especially if it is a competitor of the third party involved in the other arbitration.

Tribunals are thus more inclined to admit a previous award, or excerpts thereof, if a party voluntarily offers to produce it to support its case. Even then, the weight it will give to such evidence will still depend on its relevance to the case the tribunal has to decide, particularly in view of possible differentiating factors as noted above.

5. Concluding remarks

When it comes to evidence, gas pricing arbitrations – like other commercial arbitrations – must contend with issues of burden and standard of proof, admissibility issues, and evidentiary presumptions. And also, like other commercial arbitrations, they often involve documentary evidence, fact witnesses and expert witnesses. However, the particularities of these arbitrations bring with them also certain specificities when it comes to evidence and the best manner to support a case theory. These particularities must be navigated carefully, in order to secure the most effective outcome.

Confidentiality in gas price reviews

Philippe Pinsolle
Quinn Emanuel Urquhart & Sullivan LLP

Gas price reviews are based on a paradox: their purpose is to ensure that the price at which gas is bought matches the price at which it can be sold on a given market, yet the price at which gas can be sold on a given market (or any sub-segment of it) is among the most sensitive information even though, today, many markets are dominated by spot prices, at least in the United States, the United Kingdom and part of continental Europe. Asia and southern and eastern Europe are of course in a different situation.

Moreover, there is often an imbalance between the parties: the seller knows at which price it sells the gas to all of its buyers, while the buyer knows only at which price it buys the gas from this seller (and possibly other sellers). Conversely, the buyer knows the price at which it can sell the gas to its customers, yet generally the seller, unless it operates on the downstream market, can know about the price on the downstream market only through the information that it receives from its buyers, or possibly the regulator or other publically available sources, where the information published may lack transparency.

As a result, the level of information can vary significantly from one party to another. In addition, most gas supply contracts contain confidentiality provisions which further limit the flow of information. This limited availability of information affects the whole process of gas price reviews, including the pre-arbitration negotiations and the arbitration itself. Confidentiality also plays a role after the arbitration.

1. Confidentiality issues during pre-arbitration negotiations

Although it may not generate as much attention as is the case during the arbitration, the issue of confidentiality during the time leading up to arbitration is of major practical significance. This pre-contractual period may cover either the compulsory negotiation period before any price review arbitration, or the informal discussions which frequently are held between the parties before or during such compulsory negotiations. Practice shows that a major distinction should be made between two situations. When the parties provide for a specific agreement governing confidentiality, either in the initial contract or in a subsequent agreement, generally its terms will be enforced. The real difficulty arises when the parties do not have a specific agreement governing the confidentiality of their discussions, especially when they have differing expectations as to how they can use documents and information exchanged during these discussions in a subsequent arbitration.

1.1 When there is a specific agreement between the parties covering the negotiations

In essence, the issue of confidentiality before the arbitration is the question of whether – or to what extent – the information exchanged between the parties, and the positions that they take, can be used in a subsequent arbitration. Practice shows that some parties are aware of this difficulty and attempt to deal with it by entering into various types of agreements. Some examples will be discussed below. The effect of such agreements is that, generally, the documents and information exchanged in the pre-arbitration negotiations cannot be used in a subsequent arbitration – but much will depend on the way in which the specific agreement is drafted.

(a) The issue

Often, the parties enter into informal discussions before formally triggering the price review mechanism. In addition, most price review provisions impose a negotiation period between the parties (generally between three and six months), during which they must attempt to resolve the disputed issues and agree a new price. Most agreements provide that no party may start arbitration proceedings before the end of the mandatory negotiation period.

Typically, during this period the parties will exchange views and economic analysis on the relevant market, and state what they each think the new price should be. The buyer may ask the seller for changes in the price formula, such as a lowering of the P0, introducing spot-market indexation to the formula or modifying the weight between spot indexation and oil-based indexation. If the price review is triggered by the seller, it will tell the buyer what type of changes it requires, typically an increase based on various reasons. For this exercise to be meaningful, the parties should be free to exchange views without being bound by them in a subsequent arbitration. The aim of these informal or mandatory negotiation periods is to reach an agreement between the parties and thus avoid arbitration. As a result, the parties make concessions in order to reach a mutually satisfactory commercial agreement, notwithstanding the fact that their strict legal position may be different.

Generally, it is advisable to provide that these discussions will be conducted under a specific confidentiality regime and that the information exchanged will not be used in a subsequent arbitration. This will show that the parties understand the situation, and have taken it into account in their actions during the negotiations. It is also advisable to provide in the confidentiality agreement an automatic sanction in the event of breach (such as the impossibility for the parties and the arbitrators to make use of the evidence introduced in breach of the confidentiality agreement). A study of various gas price review agreements, and subsequent correspondence, shows that many parties often provide for a specific confidentiality regime, either in the original agreement or in a subsequent agreement – usually entered into at the beginning of the negotiations. Some examples of these provisions are discussed below.

(b) Examples of provisions

Typically, there are three types of arrangement that the parties may adopt.

The first is an agreement under which the positions expressed by the parties during the negotiation are expressed without prejudice to their rights in an arbitration – that is, in an arbitration they cannot rely on the positions expressed during the negotiation. This is not a confidentiality agreement as the information and documents are not subject to a confidentiality requirement. An example of this type of provision is the following:

> *The parties hereby recognise that these negotiations are conducted for the purpose of finding a commercial settlement to their price review dispute. As a result, the parties hereby acknowledge that the positions that they express during the course of this negotiation are made without prejudice to their rights in an arbitration.*

It is not unusual for such a clause to contain an express prohibition for the parties to rely on the position communicated during the course of the negotiation in a subsequent arbitration: "The parties undertake not to rely on the positions expressed during the course of the negotiation in a subsequent arbitration."

The second type of arrangement is a confidentiality agreement which covers all the information and documents exchanged during the negotiations. The confidentiality agreement can remain in force for an indefinite time. In that case, it can be terminated by either party with notice, which is generally fixed by the contract. Following termination, the information that had been exchanged is no longer confidential. In the alternative, the agreement can provide that the confidentiality period lasts for a specific time. The idea is that by limiting the confidentiality period, the parties will have a good incentive to reach an agreement within that time.

However, the fact that the parties entered into negotiations – as opposed to the content of the negotiations – very often is not confidential.

Generally, this second type of arrangement contains a provision which allows either party to make use in a subsequent arbitration of information that was already in their possession before the confidentiality regime. The aim is to avoid the situation in which a given document or piece of information would become confidential simply because it has been exchanged in the course of the negotiations, if it would otherwise be available. An example of such provision is:

For the avoidance of doubt, documents disclosed or exchanged by a Party in the course of the Negotiations can be used and produced by that Party in any arbitral or other legal or other proceedings.

The third type of arrangement is a variation of the second. It contains virtually the same provisions, but adds a sanction in the event that the agreement is breached. The weakness of any confidentiality agreement is that it is very difficult to enforce if a party decides to breach it. Once the breach has occurred, the information is no longer confidential and in practice the only available remedy is damages. However, in an arbitration this remedy can include excluding the record of the document containing the protected information. Of course, this will not address the fact that the tribunal might have read the document, but at least it will prevent both the party, it in its pleadings, and the tribunal, in its award, from relying on it.

As a result, many agreements provide that, in the event that the confidentiality agreement is breached, the arbitral tribunal must declare the document containing the protected information inadmissible and refuse to include it in the record and/or strike it off the record. This type of provision is best used when it is expressed as a mandate to the arbitral tribunal and, thus, as a limitation of the tribunal's procedural powers.

An example of this type of provision is:

Should a Party produce, or rely on, a document, or a witness statement, relating to discussions or (without limitation) any and all information, documents, statements, views and/or positions taken, disclosed or exchanged by or between any of the Parties in connection with the Negotiations, the Tribunal shall declare such document, or witness statement, inadmissible and strike it from the record.

(c) Practical effect of such provisions

The effect of the arrangements described above is that an arbitral tribunal will enforce their terms. Generally, there is no difficulty in enforcing the third type of agreement (ie, a confidentiality agreement which provides for the sanction in case of breach). An arbitral tribunal will have a duty to enforce it, provided that the agreement, to the extent that it is a separate agreement, makes reference to the same arbitration agreement as the one contained in the gas supply contract on which the arbitration will be based.

Similarly, in general the first type of arrangement (ie, an agreement which provides that the positions of the parties are expressed without prejudice) is also enforced without difficulty. An arbitral tribunal will not allow a party to rely on a position expressed during the negotiations. The only practical difficulty with this type of agreement is that the arbitral tribunal will be aware of the parties' positions and may be influenced by them in the course of its deliberations.

The second type of arrangement can lead to difficulties, the main one being what sanction the arbitral tribunal should adopt if the confidentiality is breached. Typically, an arbitral tribunal will decide that the documents or information submitted in breach of the agreement cannot be used in the arbitration, either by excluding it from the record or by declaring it inadmissible in the first place. However, when there is no expressed sanction to this effect, the position that the arbitral tribunal will adopt is not certain. It is recommended that a sanction covering pre-contractual negotiations is systematically included in confidentiality agreements.

Of course, if there is no confidentiality agreement at all, the difficulty is compounded.

1.2 In the absence of specific agreement between the parties covering the negotiations

If the parties have not entered into any agreement covering the status of the pre-arbitration negotiations, the question is whether there is any obstacle to the use of the content of those negotiations in the arbitration. In this regard, the parties can have differing expectations, depending on their legal background and degree of sophistication. An arbitral tribunal has to find possible solutions to this issue.

(a) The issue

If the parties have not submitted their price review discussions to a confidentiality regime, the issue is whether the content of these discussions can be used in a subsequent arbitration. This issue is twofold: first, it is necessary to determine whether the documents are admissible and, secondly, in the event that they are, what weight should be given to them by the arbitral tribunal.

This situation may become even more complicated if parties unilaterally give the documents labels such as "without prejudice", "without prejudice for commercial discussions only", "commercially confidential" or "for the sole purpose of negotiation discussions". The difference with the previous examples is that, in this case, it is often only one party that has signalled an interest in a certain protection, while the other party generally did not react.

(b) Managing the respective expectations of the parties

Often, the parties will have different expectations of the status of pre-arbitration negotiations. For a variety of reasons, sometimes they may not be able – or may not be willing – to clarify the situation by addressing the issue with the other party. In such situations, the arbitral tribunal is confronted with differing and possibly contradictory views of the parties as to the status of those negotiations. An example that the author recently encountered may help to understand this type of situation. In this case, the parties had exchanged documents in the course of pre-arbitration negotiations for the purpose of finding a new price formula. During the negotiations, one party, but not the other, had marked all its documents as confidential and without prejudice. Subsequently, the other party produced those documents in the arbitration and the first party objected. The arbitral tribunal admitted the documents in question in the arbitration. Needless to say, the party that took the precaution of earmarking its documents was disappointed by the arbitral tribunal's decision, especially because the decision did not give the reasons on which it was based.

Situations like these occur more frequently than they should. As a result, it is submitted that an arbitral tribunal should be extremely cautious when confronted with this type of situation. Several options are available to the tribunal.

(c) ***Possible solutions***

An arbitral tribunal faced with the situation in which neither party has requested confidentiality in the pre-arbitration negotiations is in a relatively easy position. In principle, neither party should object to the use of the documents or information exchanged during negotiations in the arbitration. However, should one party object, it seems that the principle should be that the document should be admitted unless, of course, there is a compelling reason why the document or information should be excluded.

It would be possible to raise an objection against the production of such a document (or the use of such information) based on the application by analogy of the rules on settlement privilege. The rationale would be that pre-arbitration negotiations are similar in nature to settlement discussions. They are intended to find an amicable solution to a given dispute. As a result, they should be excluded by nature from the arbitration, if only to protect the parties from any concession that they may have made during those negotiations.

This view, if it were to prevail, would produce rather radical effects in that it excludes from the arbitration record any documents arising from the negotiations. It seems to us that, given that parties in gas price reviews are sophisticated parties, one should not presume that they have intended to protect the information disclosed in the course of negotiations if they have said nothing to this effect. This is especially true given that tailored confidentiality agreements are in principle the norm in the gas industry.

As a result, it should be determined whether there was a common intention of the parties, expressed or implied, to submit the information exchanged during the negotiations to a confidentiality regime. Without this common intention, it should be presumed that the documents can be disclosed unless one party has in good faith attempted to protect its own documents or information. It is a question of fact, and thus it is a case-specific situation. Nonetheless, one cannot exclude the case where one party genuinely believed, or had been led to believe, that the negotiations were confidential. In that case, in our view, the documents should be excluded from the arbitration record.

Assuming that the documents are admissible, another issue is what weight should be given to them. In our view, because these documents represent an attempt by the parties to find an amicable solution, their weight should be limited. It would not be right to read in such documents concessions made by a party and for the tribunal to consider figures or proposed formulas contained in them as a ceiling (or a floor, as the case may be) to a party's claim in the arbitration. This is particularly so since the concessions may be irrelevant and misleading outside the specific context of the negotiations. For instance, the negotiations between the parties may have been broader than the mere price review and included other, unrelated issues, so that a concession on a given point was only made in consideration of an advantage on another matter, which is not to be considered by the tribunal.

For all these reasons, it is submitted that the information and to the position taken by the parties in the course of amicable settlement should be given only a limited weight.

The next issue is the issue of confidentiality during the arbitration process.

2. Confidentiality issues during the arbitration

As the arbitration progresses, the parties will have to discuss the underlying economic data to their respective claims and positions. More specifically, generally the claimant will have to show that there is a significant change in the value of gas on the relevant market in comparison with the date of the previous price review (the wording of the price review provision may vary, but in essence this is the test to be satisfied). In order to do so, both parties will have to rely on a variety of economic data, some of which is sensitive, such as pricing provisions in various contracts, which are frequently subject to confidentiality agreements. Even if they are not, one party may not want to disclose them to the other party because they are considered to be highly sensitive. In addition, there is often an imbalance between the parties as the buyer generally knows only at what price it can buy and sell gas on the end-user market while the seller, which will generally have several buyers in a given market, will often have a better picture of the price at which the gas is sold at the entry points of a given market.

As a result, confidentiality may hinder a party's ability fully to present its case or, at least, to present it in a manner capable of convincing an arbitral tribunal.

Against this background, it is useful to describe the tools that are available to the arbitral tribunal in terms of confidentiality, then to discuss the impact of confidentiality on access to information, and, finally, it is important to recall that the standard of proof must be adapted, in our view, to the availability of the information.

2.1 The tools available to the arbitral tribunal

Arbitral proceedings may be confidential because the parties have selected arbitration rules which provide such confidentiality. In addition, certain arbitration laws also provide for confidentiality. This is not specific to gas price reviews. The parties may have entered into a specific agreement creating a confidentiality regime for their arbitration. Finally, the arbitral tribunal has the power to take appropriate protective measures in order to ensure confidentiality, including recourse to a confidentiality expert.

(a) Confidentiality of the proceedings in general

Some arbitration rules provide for confidentiality. The most notable are the London Court of International Arbitration Rules, which provide as follows:

> *[U]nless the parties expressly agree in writing to the contrary, the parties undertake as a general principle to keep confidential all awards in their arbitration, together with all materials in the proceedings created for the purpose of the arbitration and all other documents produced by another party in the proceedings not otherwise in the public domain – save and to the extent that disclosure may be required of a party by legal duty, to protect or pursue a legal right or to enforce or challenge an award in bona fide legal proceedings before a state court or other judicial authority (Article 30.1).*

The International Chamber of Commerce (ICC) Rules merely state that the arbitral tribunal has the power to enact confidentiality measures. Article 22(3) of the 2012 ICC Rules provides that:

> *upon the request of any party, the arbitral tribunal may make orders concerning the confidentiality of the arbitration proceedings or of any other matters in connection with the arbitration and may take measures for protecting trade secrets and confidential information.*

The French legal regime accepts the idea that arbitration should be confidential in domestic proceedings, but remains neutral on confidentiality in international proceedings.[1]

These provisions are not specific at all to gas price reviews and are quite familiar to all international arbitration practitioners. However, they have an impact on gas price review arbitrations, as it is frequent that many cases run in parallel.

When there is a global and significant change on the market, either the sellers or the buyers will want to trigger a price review. These price reviews, depending on the relevant dates of each contract, may occur at approximately the same time. Moreover, many contracts provide for a 'joker' price review

---

1 See Article 1464(4) of the Code of Civil Procedure for domestic arbitration, which has no equivalent in international arbitration. See also P Pinsolle, "The Distinctive Features of the New French Law on Arbitration" in C Müller and A Rigozzi (eds), *New Developments in International Commercial Arbitration* (Schulthess, 2011) 99.

which can be exercised at any time and, if the change in the market is significant enough, many parties will invoke it. As a result, many cases raising exactly the same issues may run in parallel or consecutively, but more or less at the same time. As a result, different price reviews may be running in parallel at the same time in the same market under different contracts, which may deal with similar issues and be triggered by similar events.

The issue then becomes whether an award obtained in a given case – or a position taken by a party in a given case – can be used in another case (of course, to the extent that it becomes known to the other party in that other case). In practice, this may happen when several companies belonging to the same group oppose the same buyer (or the same seller, as the case may be), or when various companies are represented by the same counsel, or use the same expert, who may have access to awards rendered in parallel cases.

The provisions described above may or may not allow such use. If they do, for example, by relying on the exception according to which confidentiality does not apply when one party wants to enforce a legal right, a party may nonetheless decide to breach the confidentiality provision. Again, this discussion is not specific to price review arbitrations, but it is worth mentioning here because of its impact in this particular context.

The issue cannot be resolved in the abstract – it requires a case-by-case analysis.

(b) The impact of specific agreements entered into by the parties

A study of the contractual practice of gas supply agreements shows that some, but not all, contain confidentiality provisions applying to dispute resolution. Some of these are quite specific and will be discussed below; others are more general.

The most frequent provision concerns expert determination. Frequently, long-term supply agreements provide that certain limited questions (eg, the quality of gas) must be referred to expert determination instead of or before arbitration. In that case, many of these agreements will require the expert to be bound by a confidentiality obligation. This obligation is complete only when the expert accepts its mandate. Nonetheless, this is the most frequent type of confidentiality provision used in practice.

It is far more unusual to find a similar provision concerning arbitration. A study of 21 different contracts with a price review mechanism has not revealed any such provision. In fact, when it comes to arbitration, the parties may want to tailor the confidentiality obligations to arbitration, but in a more discrete way. As we will see below, there are many different methods to achieve this, but usually they will relate to document production or the possibility for one party to give documents supporting the case of the other party.

(c) *Case-specific protective orders and other protective measures*

The arbitral tribunal has full power to protect confidentiality of certain information if it believes that the protection is necessary. In our view, this power is part of the general power of the arbitral tribunal to conduct proceedings in a way it sees fit. The most recent ICC Rules confirm the existence of this power, but in our view that power existed before the confirmation.[2] Again, the power is not specific to gas price review disputes – frequently it is used in intellectual property disputes.

The many measures available to the tribunal include the following:

- the opportunity for one party to submit redacted documents;
- the opportunity for one party to submit redacted documents along with a non-redacted version, which will be shown to the tribunal only in the event of a dispute on the scope of the redaction;
- the creation of confidentiality 'rings' or confidentiality 'clubs' consisting of a limited number of designated individuals on each side. Typically, such individuals include the lawyers or some of them, one or more designated experts, and sometimes one or more designated client's representatives;
- the possibility for the tribunal to mandate a third party to review and confirm the accuracy of specific information; and
- any combination of the above, depending of the needs of the parties.

Although sometimes these measures can be cumbersome to implement in practice (eg, if during the hearing some witnesses have to go in and out of the hearing room depending of the subject matter discussed), generally they prove useful when there are confidentiality issues. For gas price reviews, these measures are particularly useful when discussing pricing. As mentioned above, these measures are not specific to gas price reviews. Rather, they are also used in other types of arbitration, especially arbitrations involving IP rights. Nonetheless, they can be used by, or at least be a source of inspiration for, arbitral tribunals when addressing confidentiality issues in the context of a gas price review arbitration.

A rather interesting mechanism that has been used recently is the use of a confidentiality expert. This mechanism is used primarily in investment arbitrations, where the presumption is that confidentiality does not apply, in order to protect sensitive information produced by one party. This mechanism can easily be transposed to gas price review arbitrations. In gas price review arbitration, the idea is not so much to prevent the disclosure of information to the general public, but to prevent the disclosure of information to the other party while respecting due process. The mechanism of confidentiality expert

2 See Article 22(3) of the 2012 ICC Rules.

works as follows: the arbitral tribunal defines in a procedural order a confidentiality regime distinguishing generally between two levels of confidentiality: (1) disclosure limited to the experts and a selected group of attorney on the other side, and (2) disclosure to a larger group of individuals that includes a representative of the client. In all cases, the individuals having access to the protected information must sign a confidentiality agreement beforehand. There is sometimes a third level of confidentiality: tribunal only, but this requires a specific agreement of both parties as this is a serious limitation to due process.

The confidentiality order also describes the criteria defining access to each level of protection. Once this is done, the parties are supposed to apply the order and ascribe a certain level of confidentiality to the documents produced. In the event of a dispute among the parties, it is not solved by the tribunal but by a confidentiality expert whose task is to verify that the confidentiality order has been complied with. This mechanism works well when there is a minimum level of cooperation between the parties. It has however some drawbacks: first, it is cumbersome and expensive, and secondly the nature of the mission of the confidentiality expert may vary from one jurisdiction to another. Some will take the view that the expert has no adjudicative powers and must apply the confidentiality order without, for example, having the power to interpret it. Others, rather, will view the confidentiality expert as being akin to an arbitrator on confidentiality and enjoying broader powers. Overall, this mechanism should probably be reserved to the largest cases given the costs associated with its implementation.

2.2 Access to information

In price review disputes, one of the most sensitive and difficult issues is access to information. Inevitably, parties will have to demonstrate the value of gas on a given market, and possibly the value of competing fuels. Information relating to the value of gas is commercially sensitive and players are not willing to share such core information on their business. Thus, information relating to price is difficult to access and parties must resort to external sources. Given the scarcity of information, it is even more difficult for the tribunal to determine the reliability of the information put before it. This can have an impact on a claim based on a change in the value of the gas in the relevant market. However, the issue of the burden of proof is separate from that of access to information.

As is often the case, the parties may limit or expand their ability to access the relevant information as between themselves. The arbitral tribunal will have to direct the disclosure process in a way that accommodates the legitimate need of the claimant to access the relevant information, the need of the respondent to keep confidential certain sensitive information and, possibly, the competing interest of third parties.

(a) *The distinction between burden of proof and access to information*

International arbitration practitioners are familiar with the classic objection that is raised in the context of disclosure requests. Very often, the party trying to resist a disclosure request will respond that the information requested should be produced or obtained by the other party as part of its duty to discharge its burden of proof. In other words, many parties believe that it is not legitimate for their opponent to try to obtain in their own record information which may be useful for the opponent to advance its claim. Those parties will consider that they have no duty to help the other party in this respect.

In reality, the situation is more subtle than this. There is a fundamental difference between the burden of proof and the access to information. An example may help. Assume a party has sold one of its subsidiaries to a third party, and that third party makes a number of claims for breach of representations and warranties. The selling party is aware that the records of its former subsidiary contain relevant information which may help it to resist the claim. In our view, in this particular situation, it is only right that the selling party is able to obtain, through a disclosure process, access to this information. This is not a question of burden of proof: the selling party will still have to support with evidence the fact that it asserts in its defence, but it will be given access to the relevant information, which happens to be in the possession of the other party.

Often, the same is true in the context of gas price reviews. At best, the buyer has a limited knowledge of the market's structure and the prevailing price on the wholesale gas market. This may be increased through consultants that provide a quantitative analysis of a given market. The seller may have additional information on the market because it is selling to more than one party. This may create an imbalance and this imbalance may work to the detriment of the buyer when it is the claimant in the price review process. As a result, and subject to certain conditions, the buyer should be able to obtain, through the disclosure process, access to certain information that the seller holds. Similarly, when there is an imbalance to the detriment of the seller, the seller should be able to access the relevant information held by the buyer.

However, sometimes the parties themselves limit this ability.

(b) *The impact of specific agreements entered into by the parties*

Gas price review disputes are disputes between sophisticated parties. They are aware of the fact that a tribunal has authority to order the disclosure of documents and information. As a result, they sometimes decide to limit this ability.

Our study of the practice in this regard has identified three types of provisions.

The first is to remove the power of the tribunal to order disclosure of documents, either by expressly excluding this power in the arbitration agreement or by excluding the relevant provisions of the applicable rules of arbitration (eg, Articles 24(3), 27(2) and 16(3) of the UNCITRAL Arbitration Rules 1976, insofar as such provisions relate to the production of the documents).

A second type of provision is a limitation as between the parties, but which does not specifically apply to the tribunal. It may be in the price review provision itself, rather than in the arbitration agreement, and the question then becomes whether it binds the arbitral tribunal too. Typically, this provision will be worded: "No party shall be required to disclose any business secrets nor to provide such information as the other party may need to substantiate its position."

Another provision, which is inspired by this second type of provision, imposes a duty for each party to substantiate its claim without any possibility to have access to information held by the other party:

> *Each party shall provide all necessary information to substantiate its own claim. No party shall be required to disclose any business secrets or breach any confidentiality undertaking nor to provide such information as the other party may need to substantiate its claim.*

In a third type of provision, the parties explicitly exclude information covered by a confidentiality agreement. The provision generally reads as follows:

> *During the course of any contract sales price review, or any resulting arbitration proceedings, neither buyer nor seller shall have any obligation to disclose any information which it is prohibited from disclosing by confidentiality obligations.*

(c) ***Shaping the disclosure process***

In view of the above, when considering a request for disclosure an arbitral tribunal should pay particular attention to the specific agreements entered into by the parties. Of course, this is true when the disclosure is excluded altogether. However, very often disclosure will not be completely excluded but merely limited.

The difficulty lies with the interpretation of provisions that partially limit disclosure. In particular, the issue is whether the confidentiality limitation entered into for the purpose of the negotiation will apply also during the arbitration. In our view, the arbitral tribunal should respect the terms of the confidentiality agreements entered into by the parties even though they may not explicitly state that they will apply during the arbitration. We believe that it would be counterintuitive for the parties to have submitted their negotiations to a specific confidentiality regime, but then to adopt a radically different regime for the purpose of the arbitration. If the parties have decided that they should be able to reach an agreement on the basis of a certain set of data, or on a basis of limited information, in principle the same set of data or limited information should be available for the arbitration.

The same applies if the agreement specifically states that the limitation applies during the course of the arbitration.

The existence of this limitation does not mean that the arbitral tribunal is deprived of any powers. In our view, the arbitral tribunal should direct the disclosure process in the most appropriate manner. The aim should be to limit any imbalance between the parties for access to the information, while taking into account any agreement entered into by the parties. In resolving this question, the arbitral tribunal should also take into account the limited availability of information and adjust the standard of proof accordingly.

2.3 Adjusting the standard of proof to the availability of information

There is a direct link between the standard of proof and the availability of information. The evolution of the case law on corruption illustrates this issue.

(a) The link between the standard of proof and the availability of information

International arbitral tribunals have long recognised that the standard of proof and the availability of information should be linked. The issue here is not to discuss the standard of proof in the abstract – that is, to determine whether the appropriate standard is the balance of probabilities or otherwise. Rather, the issue is to adjust this standard in real terms to the information available in a given case.

This is not specific to price review disputes, and practical solutions have been found in other fields. For instance, matters of corruption raise similar issues. For a long time, arbitral tribunals considered that, because corruption is a very serious allegation, the corresponding standard of proof must be very high.[3] The difficulty with this analysis is that, because corruption activities are clandestine by nature, it is often very difficult to satisfy such a standard of proof. As a result, the modern approach has been to adopt the opposite rule and to consider that the standard of proof applicable in corruption issues can be relatively low. The idea behind this reversal is that when sufficient indices show a pattern which can lead to the conclusion that corruption exists, this

---

3 See Award dated 19 December 1991 (Claude Reymond president) in the Westinghouse case, reference in 12(2) *ASA BULLETIN* 244 (1994); M Hwang SC and K Lim, "Corruption in Arbitration – Law and Reality" (2012) 8(1) *ASIAN INT'L ARB J*, see www.kluwerlawonline.com/abstract.php?area=Journals&id=AIAJ2012001.

4 K Mills, "Corruption and Other Illegality in the Formation and Performance of Contracts and in the Conduct of Arbitration Relating Thereto" in ICCA Congress Series No 11 288, 295 (2003). The author affirms that: "Because of the near impossibility of 'proving' corruption, where there is a reasonable indication of corruption, an appropriate way to make a determination may be to shift the burden of proof to the allegedly corrupt party to establish that the legal and good faith requirements were in fact duly met." See also C Partasides, "Proving Corruption in International Arbitration: A Balanced Standard for the Real World" (2010) 25(1) *ICSID REV* 47, 60. Partasides' proposition is that "plausible evidence of corruption, offered by the party alleging illegality, should require an adequate evidentiary showing by the party denying the allegation".

conclusion is valid unless the other party proves otherwise by providing indication or factual elements which can disprove corruption.[4] In practice, it does not constitute a complete reversal of the burden of proof, but it can amount to a burden of proof which can be satisfied on the basis of *prima facie* evidence.[5] As a result, it is a realistic recognition of the fact that because information is rarely completely available in corruption cases, the standard of proof should be adjusted accordingly.[6]

In our view, the same is true in gas price review disputes, although to a lesser extent. As explained above, the difficulty is that very often information on the value of gas is sensitive and so difficult to obtain. As a result, a party to a gas price review dispute will have to rely on internal evidence or extrapolations based on its own experience. Typically, a buyer will have direct experience of the sales that it has concluded on the end-users market and those that it has not concluded. Quite often, for the sales not concluded, the buyer will have a database reporting the content of conversations between the commercial division and the client on the level of the competing and successful offer. The reliability of such a source is not absolute, but it is an indication. This indication can be completed by quantitative studies that can be provided by some consultancies. Again, the reliability will depend on the size of the sample and on the rigour of the analysis carried out. However, because generally there is no open market that gives a reliable indication of the price, it is submitted that this type of evidence should be accepted by the tribunal and that the standard of proof should be adjusted in view of the scarcity of the information.

(b) Satisfying the standard of a burden proof in practice

It follows from the above that the burden of proof of a party triggering a gas price review can be satisfied using one or more of the following sources of information:

- public indices on the given market when they are available and representative;
- written resolutions of the regulator when they are available;
- the party's internal return or experience (realised sales and lost sales);
- any extrinsic evidence showing the price at which the sales are made in the relevant market;
- quantitative studies provided by consultancies; and

---

5 A Timothy Martin, "International Arbitration and Corruption: An Evolving Standard" (2004) 1(2) TDM 6–7. See also Final Award, ICC Case No 6497 of 1994, XXIVA YB COMM ARB 71 (1999). The arbitral tribunal stated that it: "may exceptionally request the other party to bring some counterevidence, if such task is possible and not too burdensome. If the other party does not bring such counter-evidence, the arbitral tribunal may conclude that the facts alleged are proven".

6 Partasides, fn 4 above, at 57: "in determining an appropriate standard of proof, arbitration tribunals should take account not only of the seriousness or likelihood of the allegation, but also the intrinsic difficulty of proving it."

- any other evidence that can be collected by experts in the course of preparing their report, which should be relevant for the market.

In turn, some or all of the above evidence can be confirmed by an independent third-party auditor, which can test that it corresponds to the company's record. This is particularly useful when a party relies only, or primarily, on its own internal records without any corroborating evidence. If necessary, this data can be completed by direct testimony of the representatives of the commercial division of the party triggering the price review, describing their experience in qualitative terms.

3. Confidentiality issues after the arbitration

In principle, the award decided by the arbitral tribunal is confidential between the parties after the arbitration ends. However, given the similarities between individual gas price review disputes and the contractual provisions in dispute, the access to an arbitral award may often be an advantage for the parties, if only for its interpretative value.

3.1 Limited availability of awards

The confidential nature of gas price review disputes means that is it very difficult to obtain information on the awards. Generally, arbitral awards rendered in commercial disputes are kept confidential by the parties, except when there is a duty to disclose the award (eg, to the stock market), or, at the very least, its result. The price of gas under the relevant contract is sensitive.

However, there are situations in which an award can be made available to the public. The first is when the award is challenged before domestic courts. Although it is possible in some countries to obtain a special regime because of the sensitivity of the information, generally an award challenged before a domestic court will become public, or at least partially public, through the decision of the court itself. The same is true when the award is enforced in a country other than the place of arbitration.

In our experience, awards in gas price review disputes are rarely enforced – generally, they are complied with voluntarily, often following negotiations between the parties, precisely to keep their content secret. Nonetheless, they may sometimes be challenged before the court of the place of arbitration.

Apart from this particular case, awards are usually kept secret by the parties. This is unfortunate because, provided that it is respected that the price formula is sensitive information which should be kept confidential between the parties, there are other parts of arbitral awards which may be of general interest.

3.2 The interpretative value of certain awards

Gas price review disputes are based on provisions which usually are reproduced from one contract to another. There are significant variations, but there is also language which is found in almost every contract. For instance, clauses such as "In any case, the buyer shall be able to market economically the gas" are often included in price review provisions. Information from arbitral awards that have interpreted this language would be useful for the industry at large. In addition, there are issues which arise in almost all gas price review cases, for example, what use can be made of information which becomes available after the date of the triggering of the price review? Again, access to information from awards on this particular issue would be extremely useful.

(a) The interpretation of the term 'market economically'

Most gas supply contracts contain a clause to the effect that the buyer should be able to market the gas economically on its market. Typically, this protection is granted in addition to the mechanism of the price review and may be used as a corrective device by the arbitral tribunal when determining the new price formula. The new price formula will reflect the value of the gas on the relevant market and enable the buyer to market the gas economically. Sometimes, this provision can be viewed as an independent trigger – it depends on the language of the relevant contract.

'Market the gas economically' is generally understood as enabling the buyer to sell the gas at a profit, at least on the day on which the price is reset, but not as a guaranteed margin for the future. However, there are many uncertainties as to the proper interpretation of these terms. For example, what is the proper margin? What type of costs should be taken into account?

Interestingly, this term is found regardless of the applicable law. The author has seen contracts governed by such laws as the laws of France, England and Wales, Switzerland, Italy, Germany, Denmark, the Netherlands, and New York that contain it. It is submitted that its proper interpretation is not a function of the applicable law but a function of what the industry thinks it should mean. In carrying out this exercise, it would be useful to look at what the arbitral awards have said about the meaning of 'market economically'/'market the gas economically'. For example, one possibility would be to disclose those parts of the award which do not contain sensitive information. However, to the best of our knowledge, such disclosure is not a common practice today.

(b) The issue of the date and the use of information that follows the relevant date

Another issue that is highly contested in gas price reviews is what use should be made of information on the relevant market which becomes available after the date on which the price review has been triggered. The traditional view is that when the price review is triggered at a given date, the arbitral tribunal should look at the information available on that date to determine whether the price review trigger is valid and, if so, at what level the price should be reset.

Subsequent information may be used as a benchmark or for verification purposes.

The difficulty is that, given the time taken for negotiations between the parties and then the arbitration, the award is rendered several months after this date. As a result, the arbitral tribunal will have had access to information that is more recent than that which was available on the date of the price review. In our view, arbitrators should at least take this information into consideration during the arbitration. The issue then becomes how the information can be used. One possibility is to use it as a statistical benchmark to make sure that the new formula is not way off the mark, but that is already in breach of the principles described above. Again, having access to arbitral awards on this particular issue would benefit the industry at large. One can hope that some parts of them will become available one day.

The arbitrator's role

Rahul Donde
Laurent Lévy
Lévy Kaufmann-Kohler

1. Introduction

The majority of gas sold to European (and other) buyers is sold through long-term supply contracts with 'take or pay' obligations that require the buyer to pay for an agreed volume of gas. The price of the gas sold under these contracts tracks the value of natural gas as it changes over time. If the gas is bound for a market that has a market index for natural gas, the price formula may be based on such an index. The National Balancing Point in the UK and the Henry Hub in the US are some examples. However, if the gas is destined for a market that does not have a reliable natural gas index, or if the gas is sold under an older contract that was in place before market indexation, then the gas price is normally indexed to the prices of alternative fuels, usually oil prices. These latter contracts usually contain price review clauses.[1]

As the name suggests, a price review clause allows the parties to the contract to seek a revision and adjustment of the gas price (or the price calculation formula) in specific circumstances so that the price/price formula reflects the relevant changes in the market as identified in the contract.[2] Typically, these clauses contain a procedure under which one party (or both) may notify the other of its claim for a price adjustment. The parties are then to negotiate the price for a defined period of time. If negotiations are unsuccessful, either party may initiate the specified dispute resolution process, which is usually arbitration.

A gas price arbitration is one in which an arbitral tribunal is tasked with determining whether the economics of a contract are to be adjusted, and, if so, to establish a new gas price/price formula. In effect, the parties in a gas price arbitration are asking the arbitral tribunal to operate the price review clause and set a new price or price formula, as the parties are themselves unable to do so.

While gas price arbitration is not a recent occurrence, the number, size, complexity and increasing public scrutiny of these arbitrations is notable. Indeed,

1 Ben Holland and Spencer Ashley, "Natural Gas Price Reviews: Past, Present and Future" (2012) *Journal of Energy & Natural Resources Law* 35.

2 Michael Polkinghorne and Sven-Michael Volkmer, "Price Re-openers in Long Term Gas Supply Agreements" in James Gaits (ed), *The Leading Practitioners' Guide to Oil and Gas Arbitration* (JurisNet LLC, 2015) 525.

gas price arbitrations are now commonplace.[3] They are predicted to increase significantly in the Asia-Pacific, as that region is undergoing a transition similar to that seen in the European gas markets in the past decade that triggered a significant number of gas price reviews.[4] In early 2018, an award was issued in an over US$ 100 billion gas price dispute between the state-owned gas companies of Russia and Ukraine – one of the highest-value commercial arbitrations known to date.[5] The over 1000-page award provides a good illustration of the kind of complex technical questions arbitral tribunals in these type of disputes are expected to master.[6] That award, along with others, have been thoroughly debated in the mainstream media,[7] with some clamouring for more responsible behaviour.[8] Other multi-billion dollar gas pricing arbitrations are currently ongoing.[9]

Gas price arbitrations are markedly different from 'traditional' adversarial arbitration proceedings where arbitrators are tasked with determining fault in respect of breaches of a contract or other legal wrongs. In a gas price arbitration, while legal issues may also arise, the arbitrators are to find a commercial arrangement reflective of commercial realities that is, at the same time, rooted in the parties' agreement such that the parties can continue their relationship. Technical skill and expertise is paramount – a change of just a few cents per unit of gas (or even less than a cent) can result in the transfer of millions of dollars between the parties because of the high volumes of gas delivered under the contract. Further, a tribunal's interpretation of the price review clause may resonate throughout the life of the contract, causing additional difficulties if done incorrectly or unclearly. Misunderstanding or misapplication of technical parameters may render the price formula unworkable, with the result that the parties may have to abandon the price review process altogether.

All of these factors tie in to make gas price arbitration highly sophisticated,

---

3 Stephen Anway and George von Mehren, "The Evolution of Natural Gas Price Review Arbitrations" in William Rowley (ed), *The Guide to Energy Arbitrations*, 3rd edn (GAR, 2018) 199. See https://globalarbitrationreview.com/chapter/1142622/the-evolution-of-natural-gas-price-review-arbitrations, accessed 24 February 2019.

4 Sabrina Lee, "Asia: The new battleground for gas price reviews?" (CDR, 28 June 2018), at www.cdr-news.com/categories/taiwan/8313-asia-the-new-battleground-for-gas-price-reviews, accessed 24 February 2019.

5 Douglas Thomson, "Naftogaz and Gazprom both declare win in pricing dispute" (Global Arbitration Review, 16 January 2018), at https://globalarbitrationreview.com/article/1152787/naftogaz-and-gazprom-both-declare-win-in-pricing-dispute, accessed 24 February 2019.

6 Naftogaz published the awards on its website "in response to significant public interest". See Naftogaz, "Naftogaz publishes Stockholm awards" (2018) at www.naftogaz.com/www/3/nakweben.nsf/0/0BACB9D7A7656AFBC2258355002C25B8?OpenDocument&year=2018&month=11&nt=News& (Naftogaz, 30 November 2018), accessed 24 February 2019.

7 See for instance, the reportage by the *Financial Times*, "Naftogaz of Ukrain NJSC" (*Financial Times*, 2018) at www.ft.com/stream/658a2980-46c8-4a31-887e-c135133ad80b, accessed 24 February 2019.

8 Marc-Antoine EYL-MAZZEGA, "The Gazprom-Naftogaz Stockholm Arbitration Awards: Time for Settlements and Responsible Behaviour" (Edito Energie, March 2018) at www.ifri.org/en/publications/editoriaux-de-lifri/edito-energie/gazprom-naftogaz-stockholm-arbitration-awards-time, accessed 24 February 2019.

9 Sebastien Perry, "New phase of 'gas wars' begins in Stockholm" (Global Arbitration Review, 31 January 2019) at https://globalarbitrationreview.com/article/1179842/new-phase-of-%E2%80%9Cgas-wars%E2%80%9D-begins-in-stockholm, accessed 24 February 2019.

specialised and complex. An arbitral tribunal tasked with resolving a gas price dispute is thus in unique circumstances and faced with particular challenges.

With this background, the next section reviews the 'arbitrators' role' in a gas price arbitration.

2. Arbitrators' role

The arbitrators' role in a gas pricing arbitration in essence involves interpretation and application of the relevant contractual price review provisions following the law governing the contract. The arbitrators will have to determine whether the procedural and substantive conditions for a price review have been satisfied, and, if so, how to adjust the price/price formula. While this sounds simple enough, there are, however, some important nuances. A few of these are considered below. It should be noted, however, that there is no 'standard' price review provision, arbitrators must be fully conscious of the idiosyncrasies of the parties' contract in the case before them.

2.1 Awareness of mandate

As is the case in other commercial arbitrations, in gas price arbitrations, the mandate of the arbitral tribunal depends on the terms of the arbitration agreement. Gas price renegotiation clauses typically include multi-tier clauses requiring the parties to negotiate before entering into an arbitration to determine whether gas prices should be revised, and, if so, how.

Arbitrators must pay careful attention to these pre-arbitration pre-requisites. Indeed, if these pre-requisites are couched in mandatory terms, then, subject to the applicable law, arbitrators may not immediately entertain the parties' dispute.

Consider the following example:

The request for price revision shall be made in writing and be properly substantiated by the requesting Party. Upon receipt of the above-mentioned request by the Party concerned, the Parties shall enter into negotiations within 20 days and, if an agreement is reached, sign the respective addendum to this Contract.[10]

In this (and other) cases, arbitrators would be prudent to pay careful attention to the terms of the price review clause. Subject to the specificities of each case, in some jurisdictions such as Singapore, non-compliance with mandatory preconditions to arbitration could deprive a tribunal of its jurisdiction. In other jurisdictions such as France, arbitrators would be faced with an issue of admissibility and could stay the proceedings to give the parties an opportunity to comply with the specified preconditions.[11]

---

10 Naftogaz's translation of the price review provision at issue in *Naftogaz of Ukraine v Gazprom*, SCC Arbitration No. V2014/078/080.

11 See generally George Vlavianos and Vasilis Pappas, "Multi-Tier Dispute Resolution Clauses as Jurisdictional Conditions Precedent to Arbitration" in Doak Bishop and Gordon Kaiser (eds), *The Guide to Energy Arbitrations*, 2nd edn (Law Business Research, 2017) 217.

Another facet which is perhaps unique in a gas price arbitration is that the parties may limit the mandate of the arbitrators to decide the dispute 'in accordance with' the requirements for a price review set out in the contract. These requirements may be fairly detailed, and it is for the arbitrators to follow them closely. Take for instance the following example:[12]

First, the price review clause sets out the arbitrators' remit:

(j) Any arbitration regarding a Price Review shall be made in accordance with [the price review clause]. The [arbitrators] engaged to conduct any arbitration regarding a Price Review shall be instructed to ascertain a Price Formula resulting from a Price Review in accordance with the terms and conditions of this [price review clause].

The price review clause then provides details of how the arbitrators are to "ascertain a Price Formula":

[the arbitrators] shall take into account the change, between the Reference Date and the Review Date, in the value of [the Parties'] respective Price Benchmarks [...] compared to the change, on the same dates, in the value of the Price Formula.

Any Price Benchmark shall be based on import price data available for large-scale supplies and long-term supplies of natural gas at the border of [...].

Only the following import price data [...] can be utilised for the elaboration and calculation of such Price Benchmark.

The credibility of a Price Benchmark shall be assessed [by the arbitrators] taking into account, inter alia, the relevance, significance, and reliability of the price references included in such Price Benchmark.

Evidently, the provision closely delineates the scope of the tribunal's mandate in a gas price arbitration. Parties will expect the arbitrators to follow these steps closely in setting the new price/price formula.

Other price review clauses state that arbitrators do not have the authority to amend or change any part of the contract other than the price formula. Consider the following example:

For the avoidance of doubt, any adjustment awarded shall only operate to amend the Contract Sales price and shall not extend to any other provisions of this Agreement.[13]

In these instances, arbitrators should be careful of their boundaries. In *RWE v Gazprom* for instance, despite the parties' agreement to link the contract price to oil prices, the arbitrators linked the contract price to gas spot prices.[14]

---

12 Gas supply contracts containing price review clauses are confidential. The examples referred to here (and elsewhere in this chapter) are extracts of relevant price review clauses that the authors have seen, adapted where necessary not to reveal the parties or any commercial information.

13 George von Mehren, "The Arbitrator's Role" in Mark Levy (ed), *Gas Price Arbitrations* (Globe Law and Business, 2014) 94.

14 The details of this case are not publicly available. Several commentators have however, criticised the decision for departing from the parties' bargain. See for instance, Bryan Cave, "Gas Price Reviews: Is arbitration the problem?" (2014) at www.bclplaw.com/en-US/thought-leadership/gas-price-reviews-is-arbitration-the-problem.html, accessed 24 February 2019.

Arbitrators should be careful not to alter other contract terms, however tempting it may be to do so. Provisions such as source and quality, volume flexibility, shipping expenses, diversion rights and others that have a significant effect on price but are not part of the price formula should be left untouched unless the arbitrators can root their decision in some form of parties' consent, the governing law or usages. It is, no doubt, a difficult balancing act: on one hand, these terms affect the gas price (ie, one of the 'essentia negotii', one of the core terms of the contract that are indispensable constituents of the parties' consent) and arbitrators may well have to consider and interpret them in the context of determining the appropriate gas price. On the other hand, arbitrators have to be careful not to amend these terms as they may not have the authority to do so. Parties too often complicate matters by seeking amendment of unfavourable provisions through the price review process.[15] Resolving these tricky issues often requires considerable vigilance of the arbitrators: it is for them to ensure that they fully exercise their mandate of determining the appropriate gas price, while at the same time being careful that in doing so, they do not amend the other provisions of the contract.

The same concern – of respecting the parties' bargain – applies even in those cases where the price review clause affords arbitrators some discretion in setting the gas price/price formula. In *Gas Natural v Atlantic*,[16] for instance, subject to the fulfilment of certain conditions, the arbitrators were authorised to determine "a fair and equitable revision" of the contract price. In exercising their mandate, the arbitrators introduced a two-step pricing formula, which was not requested by either party. Both parties to the arbitration eventually challenged the award.

Arbitrators must be fully aware of their mandate which is usually set out in the price review clause and must carry it out diligently. Failure to do so may well result in the award being set aside.

2.2 Compliance with formal requirements

Arbitrators in a gas price review arbitration will usually have to decide whether procedural requirements of the price review mechanism have been complied with. Common issues include the timing, form and content of the request. These issues are usually advanced by the party responding to a gas price review request, as it seeks a preliminary dismissal of the price review claim for want of compliance with these formal requirements.

The timing of the price review request is a particularly critical issue to which the arbitrators may have to pay close attention. Some sales and purchase contracts provide that price reviews can be initiated only at prescribed intervals

15 A party may, for instance, request a change of the underlying index on which the price is based. This change may, in turn, render other provisions of the contract unworkable.

16 *Gna v Atlantic LNG Co of Trinidad Tobago*, 08 Civ. 1109 (DLC) (SDNY 16 September 2008).

– for instance every four years. Others may be more specific, requiring a price review notice to be sent 'before March 28'. Still others contemplate periodic price reviews, as well as 'extraordinary' price reviews ('jokers'), which may be requested at any time depending on the occurrence of certain specified events. Subject to applicable law, if the party seeking a price review has not complied with the specified timing requirements, the arbitrators may have to dismiss the claim.

Often, the timing of the price review is linked to specified qualitative changes in the relevant market that are said to merit revision of the price formula. Take, for example, the following price review clause offered by Gazprom in all sales purchase contracts of a duration longer than four years:

The frequency and timing of the price review

The right of a Party to ask for a regular price review once in each two years based on the reasons and procedure under paragraph (19)(i)

And

The right of a Party to ask for an extraordinary price review once in each five years based on the reasons and procedure under paragraph 19(i).[17]

These 'reasons' are quite diverse and include, for instance, a 'significant change' in the economic circumstances in the European gas markets beyond the control of the parties compared to what they could have reasonably expected when entering into the contract, and/or that the prevailing contract price no longer reflects the development of gas prices at the relevant generally accepted liquid hubs in Continental Europe.

In every case, arbitrators have to be particularly careful when assessing satisfaction of given contractual requirements. Failure to fulfil the necessary requirements may require the dismissal of the price review arbitration altogether. In *Esso Exploration & Production UK Limited v Electricity Supply Board,*[18] for instance, the English High Court found a price review notice invalid as the party requesting the price review had failed to show deviation from the market price by a fixed percentage as required by the contract. As a result, the court found that no dispute had arisen that was capable of being referred to arbitration and that any arbitrators appointed would necessarily lack jurisdiction.

2.3 Different mindset

Unlike most commercial disputes referred to arbitration, in gas price arbitrations, the parties' relationship does not end at this point. Quite to the contrary, in these arbitrations, the idea is very much to maintain the long-term commercial relationship between the parties.

---

17 Commitments submitted by Gazprom to the European Commission in respect of the latter's competition concerns, available at http://ec.europa.eu/competition/antitrust/cases/g2/gazprom_commitments.pdf.

18 *Esso Exploration & Production UK Ltd v Electricity Supply Board* [2004] EWHC 723 (Comm).

Arbitrators thus need to be particularly sensitive in their conduct (and decisions) so as not to antagonise the parties. The idea is always to set a new price that is commercially acceptable and economically sustainable for both parties (although naturally one party will be less happy) so that they can continue working together. This does require a different mindset among the arbitrators: special effort should be taken to avoid approaches that are unnecessarily litigious, provocative or likely to lead to further disputes. That said, the arbitrators must remain within the terms of their remit, that is, within the agreements of the parties and the provisions of the governing law while sometimes taking into consideration usages of the industry and the earlier practice of the parties. The arbitrators must remain aware that they are not 'amiables compositeurs' although there are some contracts that do grant such authority to arbitrators.

Compounding the issue just described is another unique feature of gas pricing disputes: parties in these disputes are not necessarily seeking resolution of legal questions (of contract breach), but economic questions (of price review). Resultantly, the parties expect arbitrators to act like lawyers in these disputes but even more perhaps as 'commercial' men and women. Indeed, the only real economic difference between a gas price arbitration and a price negotiation between a gas supplier and purchaser is that parties resort to the former to reach a binding and enforceable result that they are unable to reach themselves. Parties expect, realistically or not, arbitrators to consider economic factors including market developments, prices, customs' data and other factors in the same manner as the parties would in a price negotiation between themselves.

In spite of these expectations, it remains for the arbitrators to interpret and apply the relevant provisions of the contract within the boundaries of the applicable law. While arbitrators should aim to come up with a commercially reasonable and fair price, this cannot be at the cost of the parties' actual agreement. In practice, however, this raises several questions: should arbitrators set a price that completely ignores changes in market prices post-dating the review date when the contract does not provide clear guidance on the issue? How significant must the change be for it to be a 'significant change' in the energy market of the buyer since the contract commencement date or last price review date?

The role of arbitrators in gas price arbitrations is to endeavour to step into the shoes of the parties and act as commercial men and women while still acting as judges confronted with a dispute rather than as parties to a contract. Arbitrators are to act in accordance with the law and the contract, which contract mandates them to amend the price and possibly the price clause while maintaining the parties' agreement. They are to be 'servants of several masters' or to pursue a somewhat quixotic quest, that is, to achieve a result that will make commercial sense, enabling the parties to continue their contractual relationship while respecting the parties' consent and the governing law. Failure by arbitrators to adopt a commercial mindset within the boundaries of the contract and the applicable law may lead (and has led) to unpredictable and commercially unsatisfactory outcomes.

2.4 Expertise

Arbitrators may have to address important legal issues before turning to examine whether and how to revise the gas price. For instance, arbitrators may have to decide whether a party can develop a new formula not contemplated in the notice triggering the price review mechanism. To answer this question, the tribunal may wish to turn to the negotiating history of the contract; whether they can do so would depend on the law applicable to the contract. Lengthy awards are common in gas price arbitrations, where arbitrators engage in pages-long analysis of relevant words for which dictionaries are of little help. Gas price arbitrations involving a party from an EU Member State may involve considerations of competition law. Constitutions may be relevant as well as civil codes providing for hardship. A great deal of law thus applies in gas price arbitrations.

While this is true, legal issues are, however, not always determinative of the principal issue in dispute – that is, whether to revise the gas price. The answer to this question normally includes consideration of whether a triggering event has occurred or not, and the triggers, in turn, are dependent on certain financial or economic parameters. Price adjustment criteria are based on the need to reflect the value of gas at different levels of the market, the need to ensure the marketability of prices or other factors. Key decisions in a gas price arbitration are thus commonly driven by economic evidence and market data, and less by the nuances of the law applicable to the contract or judicial precedent. While it is no doubt true that every case involves contractual interpretation, which is guided by the applicable law, it is equally true that interpretation often involves looking at words in their commercial context.

The corollary to this characteristic of gas price arbitrations is that experts, whom usually the parties rather than the tribunal will retain, are the norm rather than the exception. Arbitrators are usually confronted with a host of expert reports on a variety of issues, from developments in the gas markets to developments on a regulatory level to the value of gas in a particular market. Financial experts and industry experts opine on the appropriate economic risk allocation in practice of gas supply agreements in general, and reports setting out complex calculations on pricing, value and profitability are common.

Further, a good part of the legal submissions in these arbitrations usually corresponds to the expert evidence. Although they also address contractual interpretation questions, the real focus of the legal submissions is on presenting the economic evidence in a cogent fashion. These submissions will thus explain whether the relevant criteria for a price review have been met and, if so, the mode and level of adjustment to be made. As the parties' experts address these issues, legal submissions largely focus on the evidence contained in the expert reports.

The resolution of gas pricing disputes, therefore, turns on the arbitrators' ability to manage, understand and assess technical evidence to ensure that the parties' legal and commercial criteria have been met. Arbitrators need to have a firm grasp over the fundamentals of gas pricing as well as the role they are being asked to perform.

2.5 Economic evidence in its proper context

Another issue requiring equal vigilance from arbitrators stems from the very evidence-intensive nature of gas price arbitrations described above. The parties' experts usually diverge significantly based on the differing instructions they may receive and on the data and methodologies they find to be best suited to the facts. Often, the dispute before the arbitrators becomes a debate about which methodology or price data the price review clause directs the tribunal to use to decide various issues.

Arbitrators must be careful in these circumstances not to allow their analysis to be tainted by the volumes of economic evidence presented by the parties. Take, for example, an arbitral tribunal tasked with interpreting the words 'comparable products' in a price review clause. Parties may themselves differ significantly on whether the term should be interpreted on the basis of the 'comparable products' at the time of the parties' agreement or at the review date, or both. Arbitrators may be tempted to decide one way or the other depending on the expert evidence presented to them. They might decide that 'comparable products' must refer only to the fuels with which the relevant gas competed when the parties agreed the contract, not the fuels with which it will compete at each future review date. In doing so, however, they may miss other indicators in the contract that point to the parties' agreement to look afresh at the basket of competing fuels each time the parties ask for a review. Likewise, what should the arbitrators do if a comparator that existed at the time of the contract, and which the contract expressly refers to, subsequently disappears?

Arbitrators must be careful to ensure that their interpretation of the contract is legally sound, even in those circumstances where cogent evidence is presented to them to the contrary. Strictly speaking, arbitrators should interpret the contract in accordance with the law applicable to the contract and not the technical evidence presented by the parties, unless the applicable law itself allows for consideration of such evidence. Arbitrators should consider interpreting price review clauses in such a way that enables those clauses to function in a way that the parties intended, usually consistent with the economic or market conditions at any particular time and not fully dependent on such conditions if the parties' agreement did not consider them.

The discussion above evidences the sensitive push and pull seen in a gas price arbitration. On the one hand, arbitrators are expected to act as commercial men and women and give serious consideration and weight to the voluminous economic data presented to them, while on the other they are to remain vigilant and not allow the tail to wag the dog. Economic evidence must be considered and given weight in its proper context and must not control contractual interpretation.

2.6 Knock-on effects

Price review clauses are usually not 'single use' clauses; typically they usually allow for multiple price reviews at specified times or on the occurrence of specific events throughout the life of the contract. Put differently, parties may repeatedly submit the same core issue to an arbitral tribunal for different periods in the course of their relationship: seeking repeatedly to determine whether and how the contract price should change. Sometimes both parties simultaneously request adjustments in opposing directions: indeed, contracts may allow a party to file a request to adjust prices within a certain period running from an adverse request to adjust prices. In deciding whether and how the contract price should change, every arbitral tribunal in a gas price arbitration will follow more or less the same steps: (i) interpret the price review clause; (ii) apply the relevant facts to that interpretation (ie, decide whether circumstances exist requiring a price review); (iii) set the new price/price formula; and (iv) direct back payments if necessary.

This feature raises particular questions of estoppel in gas price arbitrations, which are usually not seen in other commercial arbitrations involving allegations of breach of contract. Indeed, under English law (and other legal systems), "limits must be placed on the rights [to] reopen disputes".[19] Therefore, "if an issue has been distinctly raised and decided [...] it is unjust and unreasonable to permit the same issue to be re-litigated afresh between the same parties or persons claiming under them".[20]

In a gas price arbitration, an arbitral tribunal may have to (repeatedly) interpret the word 'significant', where the price review clause requires a change or a difference to be 'significant' before a price review can occur. In doing so, it might define the word for future price reviews, without actually intending to do so or without fully understanding the implications of its decision. Another relevant issue is the determination of the trigger conditions of a price review as at the review date. Depending on the trigger event mentioned in the price review clause, in the first price review arbitration, an arbitral tribunal may have to determine whether there has been a "significant change in the energy

---

19 Lord Wilberforce in *The Ampthill Peerage Case* [1977] AC 547 at 569.
20 Maugham LC in *New Brunswick* [1939] AC 1 at 19–20.

market" as at the review date. Arguably, that determination of the state of the market will be the starting point for assessing a change in the next price review. In a subsequent price review therefore, should a party be able to reopen the issue of the state of the relevant energy market as at the review date for the first price review? The answer may depend on the law governing the conditions and effects of *res judicata* and it is well known that laws will considerably vary in this regard. One view is that issue estoppel should extend to the state of the relevant energy market as at the review date for the first price review, but that need not always be the case. This issue is not limited to price review clauses that require examination of the relevant energy market. Similar issues will arise in any price review clause that requires a comparison between the market at two different times, two prices at the same time (eg, the contract price and the market price or average import price on the review date) and others. A decision on any change obviously involves two factual assessments, which, strictly speaking, may prevent a future redetermination in a future arbitration.

Irrespective of the legal effects of a prior decision in the same contract, such a decision on the interpretation of a price review clause will have persuasive effects at the least and may well resonate throughout the life of the contract. Arbitrators must be sensitive to the long-term effects of the issues presented to them and should be careful not to make any determinations that may affect subsequent price reviews, unless it is necessary to do so to resolve the issues in dispute. While in some instances these long-term effects cannot be avoided, it is for the arbitrators to be vigilant and carefully determine (and limit) issues that are likely to affect future price reviews.

3. Conclusion

Gas price arbitrations are easily among the highest-value, most complex arbitrations in the universe of commercial arbitration. Billions of dollars are often at stake, with a price change of just a few cents resulting in the transfer of millions of dollars from one party to the other over the life of the contract or until the next price review. By asking arbitrators to stand in for the parties and find a commercially acceptable solution preserving the parties' long-term relationship, the parties essentially require the arbitrators to perform a role that is different to the role that the arbitrators normally perform. Arbitrators deciding these disputes therefore have a weighty and difficult task that presents some specific challenges. They must be sensitive towards the sophisticated, technical and complex nature of a gas pricing arbitration and fulfill their mandate accordingly.

The client's perspective

Nikolas Hübschen
Uniper Global Commodities SE

1. Introduction

This chapter will look at several topics related to gas price reviews from an in-house perspective. Given that price review clauses in long-term gas supply contracts (LTCs), the setting of each price review and therefore also the background of each gas price arbitration all differ, the following is necessarily a general overview of the complex and wide-ranging issues in this area and cannot replace a thorough analysis of the individual case at hand.

Before we turn to some specific aspects when preparing for and conducting a gas price review, we look at the reasons why the parties to an LTC actually agreed in most cases on price review clauses in the first place, and therefore to the often laborious exercise of price reviews. Afterwards, the chapter will address the questions whether gas price reviews are still required, and whether there are viable alternatives to arbitration for solving gas pricing disputes.

2. Why price reviews?

In North-West Europe the existence of LTCs for the supply of natural gas is closely linked to the discovery of large natural gas reserves, first in the Netherlands and afterwards also in the North Sea and Russia. The development and exploration of these gas resources required substantial investments and often external financing. Long-term off-take contracts provided the continuous cash flows necessary for the financing by virtue of so-called take-or-pay commitments by the buyers.[1] Under such a take-or-pay obligation, a buyer is obliged to either take and pay for the gas made available by the seller or to pay for a defined minimum quantity of gas, even if the buyer does not take it. Consequently, the seller will receive continuous cash flows irrespective of the off-take of the buyer. Such a take-or-pay commitment, which can easily translate into an annual payment obligation above a billion Euro for larger contracts, is obviously a substantial risk for a buyer, in particular when taking into account that the buyer usually does not consume the gas itself but needs

1 However, new gas production projects developed during recent years still rely in many cases on long-term contracts with take-or-pay obligations to secure financing.

to resell it in its downstream market. A buyer can therefore only take the inherent risk of a take-or-pay commitment if it has a kind of assurance that it will be able to resell the gas and earn an adequate margin for its risks and efforts. This requires two things from a buyer's perspective: a purchase price under the LTC which is competitive in the buyer's market at the time the contract is signed; and a contractual mechanism to ensure that the price will continue to reflect the market of the buyer over the duration of the contract.

Therefore, the price at the outset will generally be negotiated between the seller and the buyer in a way that the buyer will be able to market the gas in its market with a reasonable profit, but that at the same time the lion's share of the market value of the gas will be passed on to the seller in view of its large up-front investments. To make sure that this bargain is maintained and that the contract price follows the market as it develops, the price is then indexed to an appropriate benchmark which mirrors the pricing method, and therefore tracks the development of the overall price level for gas in in the relevant market.[2] Provided the benchmark to which the price is indexed accurately reflects the price structure and level in the relevant market, the contract price will over time automatically move upwards and downwards in line with that market. The indexation of gas prices in the LTCs thus allows the seller to maximise its returns on its investments while it also gives comfort to the buyer that the price will by and large remain competitive in its downstream market.

However, the indexation of prices helps to maintain the economic equilibrium only as long as the chosen price formula and its indexation accurately reflect the relevant market developments. A multitude of factors, for example changes in the legal framework or changes in the purchasing behaviour of customers in the buyer's downstream market, can lead to situations where the chosen price formula and its indexation no longer correspond to the market value of gas. For these kinds of developments, sellers and buyers agreed on price review clauses. Consequently, even though the wording and exact requirements of individual price review clauses differ, their fundamental purpose was (and still is) to uphold the original bargain, that is, ensuring that on the one hand the price remains market-reflective to safeguard the buyer against the take-or-pay risk[3] and allowing it to earn a net margin when reselling the gas in its market, while on the other hand that, under the changed market circumstances and given the protections afforded to the buyer, the seller gets the highest possible return for its investments.

The interplay between the take-or-pay commitment on the buyer's side and a pricing and price review mechanism to keep the price competitive in the

2 Some of the LTCs in the early days of the natural gas era in North-West Europe still contained fixed prices but since the 1970s the indexation of prices became the norm.

3 Without a take-or-pay obligation a buyer would be much less concerned about the contract price: if the gas under an LTC became too expensive to sell in its market, the buyer would simply not offtake the gas.

buyer's market has been summarised as follows: the seller under an LTC takes the price risk, the buyer the volume risk.[4] Correspondingly, even though price review clauses generally do not distinguish between a price review requested by a buyer and one requested by a seller, the underlying commercial motivations of a seller and a buyer for requesting a price review are quite different. On the one hand, a seller is mainly concerned about the price level, making sure that it maximises its revenues in view of its investments; on the other hand, a buyer is not primarily concerned about the absolute price level but rather about aligning the price with its market, that is, that the price under the LTC is and remains market-reflective and allows the buyer to cover its costs and to retain a reasonable profit.

3. Price reviews under LNG contracts?

Although the above section primarily refers to LTCs for piped gas, similar concepts can apply to LTCs for the delivery of liquefied natural gas (LNG). However, for LNG one needs to distinguish between two types of contracts: contracts under which the seller delivers the LNG to the buyer at one or more specified regasification terminals (so-called DES contracts[5]), and contracts under which the LNG is delivered at the liquefaction terminal prior to shipment (so-called FOB contracts[6]). The latter allows the buyer to ship the LNG to destinations of its choice – that is, the buyer determines in which market it will sell the LNG/gas. Hence, even though FOB contracts usually contain take-or-pay commitments, the buyer can to a certain degree mitigate the resulting risk through the possibility of delivering the LNG to different markets. Therefore, some of these contracts do not provide for price review clauses and their prices are often not based on any particular target market but on the costs of the production of the LNG and the cost for the so-called feed gas, which is liquefied. In contrast, under a DES contract the buyer is in a similar position as under a supply contract for pipeline gas. Unless it has the chance to reload the LNG onto another vessel and bring it to another market, the buyer has to resell the regasified LNG in the market downstream of the terminal to which the LNG has been delivered. As a consequence, DES contracts are usually priced in a way that reflects the pricing in such downstream market to allow the buyer to profitably resell the gas, and they regularly contain price review clauses to ensure that this principle is maintained over the lifetime of the contract.

---

4 See, eg, Energy Charter Secretariat, *Putting a Price on Energy – International Pricing Mechanisms for Oil and Gas* (Brussels, 2007) 153.

5 Named according to the previous ICC incoterm DES (delivered ex ship); even though the 2010 incoterms have abolished DES, the DES term and concept is still used, even in new contracts.

6 Named according to the ICC incoterm FOB (free on board).

4. Conducting a price review

After clarifying why the parties to an LTC actually undergo the potentially cumbersome procedure of a price review, the following sections will look at the typical price review process itself. This process can be divided into three stages: triggering a price review; the subsequent negotiation phase; and – if no solution is reached – the gas price arbitration. In addition, there will in practice often be commercial settlement discussions in parallel to the formal price review process.

4.1 Trigger phase

Most contracts will provide for specific time intervals for regular price reviews (most often two or three years) but also often include a limited number of so-called 'jokers' or 'wild cards' for each party – that is, price reviews that may be triggered during the regular intervals. The requesting party should carefully observe the formal requirements defined in the contract clause for triggering a price review, namely format, timing and required content for a valid review notice.

- Format – Most LTCs will specify that the request must be made in writing. Often the price review clause will also determine the permitted communication means. Many LTCs require letter or courier service but often do not provide for a submission by telefax or email. If one wants to use any of those 'modern' technologies, it should be carefully reviewed in each case whether these forms of communication will also be sufficient. In case of doubt, an agreement with the other party on the permissible forms of communication prior to submitting a review notice is advisable.
- Timing – The contract will usually also define a date by which the review notice must be received (in the required format and by a permissible means of communication) by the other party in order to be effective. Missing such a date can lead to the invalidity of the request. In practice, parties often agree to postpone the date when a review notice must be submitted in order to continue their commercial discussions.[7]
- Content – The majority of LTCs will require that a request identify the reasons for such a request in order to enable the other party to evaluate whether the contractual requirements for a price review are met, and that the requested revisions to the price are specified or at least sufficiently outlined. Unless exceptionally required by the contractual price review clause, a review notice does not have to contain any evidence supporting the reasons mentioned for the request. Whether or not the reasons actually exist and support the requested amendments is a matter of the merit of the claim, but as a rule does not affect the validity of the request.

---

7 See section 4.2(b) below.

The effort required for preparing a price review should not be underestimated. As a first step, a thorough legal analysis of the relevant price review clause is needed to determine what the specific formal and substantive requirements for a price review under this LTC are. Then, based on the legal analysis, relevant trigger criteria need to be established. This will require a detailed assessment of the changes in, and the present conditions of, the relevant market and the implications for the pricing under the LTC. Lastly, it needs to be determined which types of amendments to the pricing are justified and/or which ones should be demanded. In the end, the decision about the content of the specific price adjustment request depends on economic and legal but also strategic considerations. Even though a price adjustment will often not be a purely mathematical exercise with only one definite possible result, clearly exaggerated claims should be avoided. An indefensible claim will not be taken seriously, neither by the other party nor by a tribunal in case of arbitration. In addition, even if – and this will depend on the applicable price review clause – the claim can be amended during the course of the review process, a complete overhaul of the claim is unlikely to lend credibility to the case. Therefore, the party requesting a price review should invest enough time and resources into the diligent preparation of its trigger notice to avoid unnecessary challenges later.

4.2 Negotiation phase

(a) Formal negotiations

If a price request has been submitted, most LTCs provide that the parties shall meet and discuss the request and, if the request is justified, agree on an adjustment. These negotiations are often referred to as 'formal negotiations' because they are part of the formalised price review process. In practice such formal discussions these days rarely result in a solution for the price request. One of the reasons for the low success rate may be the fact that these discussions are often considered to be 'with prejudice', that is, documents and arguments which are exchanged can be used in any subsequent arbitration. Therefore, parties are often reluctant to make any concessions in these discussions. Nevertheless, the formal discussion phase is not useless. On the one hand it usually goes along with a 'peace period' (often three to six months) during which generally neither party is entitled to initiate arbitration proceedings, which can help a party to 'cool off' and reconsider its own case, weighing its strengths and weaknesses. On the other hand, it allows parties to hold in parallel so-called commercial negotiations.

(b) Commercial negotiations

Commercial negotiations are discussions which take place outside of a contractual price review procedure. In practice, commercial discussions often happen even before a price review is triggered and they also often continue even after an arbitration proceeding is initiated.

To enable an open discussion, the parties usually enter into a confidentiality agreement, which typically also specifies that the discussions and their content shall be 'without prejudice' to the legal position of the parties. The exact scope and operation of such clauses depend on the individual agreement, which therefore requires careful drafting.

The advantage of negotiations outside the boundaries of a formal price review is in particular that parties can create a solution space which goes beyond the normal scope of a price review clause:

- Most review clauses in LTCs are limited to an adjustment of the contract price.[8] However, changing the price will usually be to the advantage of one party but at the same time to the direct disadvantage of the other, which makes amicable solutions difficult.
- Including other contractual elements or even topics outside of the existing contract can make compromises on the price easier. Ideally, such solutions create a win-win-situation for both parties. This can in particular be achieved with elements which the parties value differently. For example, if the producer for whatever reason has spare production capacity, it may have an interest in expanding the delivery volume under the LTC because the additional quantities reduce the relative share of its fixed production costs per unit of gas. At the same time, the buyer may be interested in obtaining additional volumes if it is able to profitably market them at the (revised) contract price.
- Other elements that could be included in a commercial discussion comprise offtake flexibility, delivery points, payment terms, technical parameters but also matters relating to other existing transactions between the parties or new projects.
- As a measure of last resort, the parties could also agree to terminate the existing LTC, potentially combined with a compensation payment in one or the other direction. Reportedly, this has been the solution for some parties during the difficult price reviews occurring after the market liberalisation seen in North-West Europe.

While formal price reviews (and even contract reviews, if they are permitted under the LTC) always have to stay within the boundaries of the existing contract, the parties are always free in commercial discussions to re-negotiate their existing contractual relationship, even if this changes the original bargain

8 Some LTCs contain review clauses which are drafted in a way that they not only allow for price revisions but, according to their wording, revisions of the entire contract.

fundamentally. In practice, the opportunity to find solutions beyond a pure price adjustment often tremendously facilitates amicable settlements.

4.3 Arbitration phase

If the discussions do not progress (or even fail) and the agreed peace period, if any, has lapsed, the party requesting the price review can initiate the contractually agreed dispute resolution mechanism. In the vast majority of LTCs, the parties have agreed on arbitration as the dispute resolution method.

The decision to initiate an arbitration proceeding should not be taken lightly. Once the proceedings have been started by a party, under many jurisdictions they cannot be unilaterally terminated without the risk of the claimant losing its underlying claim. Therefore, if the process is commenced, the claimant should be willing to continue it until a final award is made, mindful of the necessary (and potentially significant) efforts, costs, time and uncertainties involved.

When preparing for the arbitration proceedings, parties need to take important decisions on the selection of external counsel and experts and, later on, on the nomination of arbitrators.

(a) Selecting external counsel

Even companies with extensive in-house price review experience generally use the support of specialised external law firms. Qualifications to consider in the selection process include, amongst others:

- Gas pricing experience – Gas pricing disputes are a unique area which has, in particular in recent years, seen quite a development. Therefore, it is important to have lawyers who are not only familiar with (international) arbitration proceedings in general, but also know the characteristics and the current state of play of gas pricing disputes.
- Commercial grasp – Today gas pricing disputes deal with complex commercial concepts. The ability of external counsel to understand and to explain them to a tribunal in written submissions as well as during an oral hearing will be essential.
- Resources in relevant jurisdictions – In many gas price arbitrations the substantive law of the contract and the seat of the arbitration belong to different jurisdictions. The chosen law firm should be capable of covering these.
- Availability – Last but not least the members of the team of the law firm working on the case should be sufficiently available during the entire period of the proceeding. While this may sound self-evident, the tremendous case load of the specialised lawyers in recent years warrants a consideration of this issue.

Regarding the question of when one should appoint external counsel, there appears to be no right or wrong answer. For a less experienced in-house team, it may be advisable to obtain external legal advice early in the process, that is, even before triggering the formal review process, in order to ensure a proper interpretation of the price review clause and, consequently, a realistic expectation regarding the price review process and its potential outcome. Yet even more experienced parties should not wait until the last minute, but should instead involve external counsel – at the latest – when an arbitration procedure becomes a realistic possibility in order to allow for a proper preparation. In any event, it is advisable to check early on with potential law firms whether any conflicts exist to avoid any surprises when time runs short; given that the number of players in the gas industry and of specialised law firms are both relatively small, conflicts arise more often than in other areas.

(b) Selecting experts and witnesses

Unlike state courts in many jurisdictions, arbitral tribunals will generally not appoint their own experts but will rely on experts appointed by the parties. Having qualified experts on board who support the respective positions of the parties is therefore crucial. Some aspects for selecting an expert are:

- Gas pricing experience – The chosen expert should not only be a good economist but should also have gas pricing experience and be familiar with the various methods and approaches applied in such proceedings. In addition, the expert should be capable of performing all economic analyses which are relevant for the case. This will usually require that the expert has sufficient available resources within his or her organisation.
- Experience in the relevant geographic market – Gas markets vary considerably. Therefore, the expert should be familiar with the regulatory framework, structure and practice of the market(s) relevant for the price review.
- Experience in testifying – Another key qualification of an expert is the ability – in written reports as well as in oral hearings – to explain complex commercial models and calculations to a tribunal which is typically not composed of economists. In addition, the expert needs to be able to defend his or her position in cross-examination. Hence, experience in writing reports and testifying in front of a tribunal is indispensable.
- Independence – A good expert will be and remain independent and will also challenge your case and thereby help develop it. While it may seem attractive at first, a 'hired gun' will, in reality, not assist but will, rather, undermine the case. A tribunal will, in particular with the support of the other party's expert, usually detect unreasonable positions and may then entirely disregard such expert opinion.

If a party cannot find one expert who is capable of covering all relevant aspects of the case, the appointment of more than one expert might be a solution.

Ideally, the expert will be appointed early in the process so that he/she can be involved in the development of the adjustment claim. Defining a claim first and only later asking an external expert to support it, can – depending on the claim – be challenging.

In many cases, fact witnesses may also be necessary. Unlike experts, witnesses do not give an opinion based on their expertise on the subject but testify about matters of their own experience and perception. Often witnesses are employees of one of the parties who are able to explain certain areas of the business or testify on their view on past contract negotiations. But witnesses can also be third parties who, for example, testify on their experience on certain aspects of the market. While employees are obviously paid by the respective party, one needs to be careful with remunerating external witnesses for their testimony; expense allowances and, in the case of consultants, compensation for lost income should be acceptable in principle but other payments may call the independence of the witness, and therefore the credibility of the testimony, into question. Creating transparency about any compensation granted to the witness, for example by mentioning it in the witness statement, can in some cases be a way to avoid negative consequences later.

(c) Selecting arbitrators

Most LTCs provide for an arbitral tribunal composed of three arbitrators; each party has the right to nominate one arbitrator. The ability to participate in the selection of the persons who decide on the case is commonly considered as one of the main advantages of arbitration over litigation. Accordingly, this opportunity should be exercised with great care.

Many companies and law firms will have some kind of list of potential arbitrators as a starting point. Gas pricing experience can certainly be an advantage but with the large number of gas pricing arbitrations in recent years it has become increasingly difficult to find potential arbitrators with such experience who are available and do not have a conflict. So, gas pricing experience may not be indispensable as long as the candidate has a strong commercial understanding.

One way to gain a better insight into the previous work and experience of a candidate is to interview him or her. Subject to domestic arbitration laws that are more restrictive, this is generally considered to be acceptable as long as the interview relates only to expertise, experience, ability, availability, willingness and the existence of potential conflicts of interest of the candidate, and does not address the merits of the dispute or legal arguments that may arise in the case at hand. The respective publications by the International Bar Association

(IBA) and the Chartered Institute of Arbitrators (CIArb) provide useful guidance in this respect.[9]

The appointing party and the potential nominee should diligently assess potential conflicts of interest. Even though not binding in most jurisdictions, the IBA "Guidelines on Conflicts of Interest in International Arbitration" provide a standard for this topic which is widely used in practice. Since some circumstances can be considered more problematic if not disclosed by the candidate, transparency about potential issues right from the beginning is a sensible approach.

Moreover, when selecting the party-nominated arbitrator it is important to keep in mind that there will of course be another arbitrator appointed by the other party and a third, presiding, arbitrator. Therefore, one should consider potential interactions of the party-appointed arbitrator with other members of the tribunal, in particular any chairperson. If the appointment procedure of the chairperson gives the parties a say in the appointment of the chair, the parties should already from the beginning think about potential candidates for that position and probably set up a separate list with them to prepare for later discussion with the other party. However, one should avoid interviewing these candidates as the unilateral contact with a prospective chairperson is in principle considered inappropriate.[10]

(d) Role of in-house counsel

In-house counsel play a key role in a successful gas price arbitration. On the one hand, they have an in-depth understanding of the market underlying the case and know the LTC and its clauses as well as the contract's history. On the other hand, they are able to translate this into a legal case. Therefore, in-house counsel could be described as the transmission belt between the internal stakeholders of a party, like account managers, commercial experts and ultimately the board, and the external stakeholders in the case, that is, external experts, external counsel and ultimately the tribunal.

Due to this central role, in-house counsel should also actively participate in the discussions about the strategy for the case. After all, gas price arbitrations are legal proceedings, even if they have commercial matters at their core.

In addition, in-house counsel also have various organisational tasks in the process. They need to select (potentially in cooperation with other internal stakeholders) external advisers, steer the work of the advisers, and organise the process of preparing and aligning legal submissions, expert reports and witness statements.

---

9 See IBA "Guidelines on Party Representation in International Arbitration", Rule 8(a); CIArb Practice Guideline 16: "The Interviewing of Prospective Arbitrators".

10 See IBA "Guidelines on Party Representation in International Arbitration", Rule 8(c), which provides that only "if the Parties are in agreement that such a communication is permissible", can communication by one party with a prospective Presiding Arbitrator be appropriate.

(e) *Arbitration proceedings*

Other chapters of this book deal with individual parts of the arbitration proceedings in great detail. Therefore, only two practical aspects shall be addressed here: first, contrary to the widespread assumption, arbitration proceedings, their content and the awards are, according to many jurisdictions and rules of arbitration institutions,[11] not *per se* necessarily confidential.[12] As a consequence, parties may want to agree on a confidentiality undertaking or ask the tribunal to include such a provision in a procedural order. Secondly, if internal costs for the arbitration proceedings shall be part of the later cost submissions, even though not required, it may be useful if in-house counsel and in-house commercial experts involved in the arbitration procedure fill out time sheets during the proceedings.

(f) *Post-award negotiations*

In some cases, the price review process does not end with a final award. For example, parties may find themselves in the position that neither party is satisfied with the award. In such cases parties may decide not to apply the award but instead negotiate a different settlement. In such a situation the reasoning of the tribunal may nonetheless provide helpful guidance for the parties.

In some instances a party also triggers a second price review before the first one has been settled by negotiations or by arbitration. How to deal with cases of parallel price reviews in arbitration needs to be determined on an individual basis. In many cases the most reasonable approach will be to stay the second price review process until the first has been decided in order to have the new contract price as a basis for the second review. If the first price review arbitration is still in an early stage, a consolidation of the proceedings may be an option; however, the individual price requests and their prerequisites need to be kept separately in such cases because the consolidation does by itself not affect the substantive legal position.

5. Need for price reviews in the future?

Following the transition of most LTCs in North-West Europe from oil-based pricing to a pricing methodology based on market prices for gas, there has been a discussion as to whether there will still be a need for price reviews in the future. In fact, in a number of recent new gas exploration projects based on LTCs, there has been a debate whether such new contracts should contain a price review clause at all. The main argument put forward for this position is

---

11 For example, the ICC regularly publishes some (anonymised but still often attributable to cases/parties) awards in its bulletins, usually a few years after the awards have been rendered. See also the ICC's recent "Note to Parties and Arbitral Tribunals on the Conduct of the Arbitration under the ICC Rules of Arbitration", 1 January 2019, paras 34 *et seq.*

12 See also Philippe Pinsolle's chapter, "Confidentiality in gas price reviews".

that 'the market' will find the right answers for all future developments and will reflect them in the market price. However, this view entirely disregards the fact that most gas markets have become subject to intensive regulation with potentially distorting effects. In addition, as the past decades have impressively demonstrated, it is impossible to foresee or even simply guess what developments will affect the gas markets during the (frequently long) duration of new LTCs. In fact, practice shows that changes requiring a price adjustment can even occur in the shorter term, for example because the applicable regulatory market regime or the gas purchasing behaviour of customers in a given market changes in a short period of time. Therefore, price reviews continue to occur regularly even under LTCs which have been adjusted to a gas-indexed price. Against this background, it generally appears sensible to uphold price review provisions in existing contracts and to include such clauses in new LTCs as well.[13]

6. Alternatives to gas price arbitrations?

The large number of gas price arbitrations and the resulting costs, the uncertainties during the often time-consuming proceedings and the sometimes unexpected results have led to a discussion in the gas industry on whether there are viable, or even more suitable alternatives to solving gas pricing disputes by arbitration. In general, three potential alternatives come to mind: litigation in front of state courts, expert determination[14] and mediation.[15] The pros and cons of the individual dispute resolution methods for gas pricing issues are summarised in Figure 1.[16]

As can be seen from Figure 1, in the area of gas pricing disputes litigation seems to provide almost no benefits compared to arbitration. While in certain jurisdictions it may be less costly than arbitration, inter alia the generally missing gas pricing expertise by courts and the difficulties of enforcing state court decisions in foreign countries (due to the lack of an international law instrument similar to the New York Convention[17]), does not make litigation a realistic alternative to arbitration.

Expert determination certainly has its merits in more technical disputes. However, even though many gas pricing disputes have a commercial rather than

13 For advice on drafting such clauses see Mark Levy's chapter, "Drafting an effective price review clause".

14 See James Farrell, "Expert determination" in *Dispute Resolution in the Energy Sector: A Practitioner's Handbook* (Globe Law and Business, London, 2012) 39 *et seq* for a more detailed analysis of expert determination proceedings.

15 See Georgia Quick and Deborah Tomkinson, "Alternative dispute resolution" in *Dispute Resolution in the Energy Sector: A Practitioner's Handbook* (Globe Law and Business, London, 2012) 51 *et seq* for more details on mediation and other forms of ADR.

16 The table is a simplified overview of the general advantages and disadvantages. For example, as mentioned above, confidentiality is not necessarily an automatic feature of arbitration, but parties can, like for mediation and expert determination, agree on confidentiality, which will in most jurisdictions be much more difficult for litigation.

17 Convention on the Recognition and Enforcement of Foreign Arbitral Awards (New York, 1958).

Figure 1. Overview of the general characteristics of dispute resolution mechanisms

	Arbitration	Litigation	Expert determination	Mediation
Confidential?	✔	✘	✔	✔
Fast?	✘	✘	✔	✔
Cost-effective?	✘	✘	✔	✔
Enforceable?	✔	✔	✘	✘
Industry expertise?	✔	✘	✔	✔
Legal assessment?	✔	✔	✘	✘
Commercial assessment?	✘	✘	✔	✔
Parties control the result?	✘	✘	✘	✔

a legal question at their core, they will in most cases also include legal issues, namely the proper interpretation of the price review clause. In addition, expert decisions are not themselves directly enforceable. Therefore, expert determination does not appear to be a viable alternative (although it is sometimes seen).

As Figure 1 shows, mediation has many advantages. In fact, it has become more widely used in gas pricing disputes either prior to initiating arbitration or also in parallel to arbitration proceedings in recent years. However, it is an instrument to support commercial discussions rather than to solve a formal dispute because it ultimately requires the consent of both parties. However, a viable dispute resolution mechanism will require a decision by a third party if one contractual party rejects any amicable solution. In practice the mediation of gas pricing disputes is also often impeded by the lack of experienced mediators in this area.

Thus, although arbitration certainly has its shortcomings, it still appears as the best ultimate dispute resolution mechanism or – to paraphrase a famous quote by Winston Churchill: "Arbitration is the worst form of dispute resolution for gas pricing, except for all the others."

The opinions expressed in this chapter are those of the author and not of Uniper Global Commodities SE or of any of its affiliates. The examples used in this chapter were chosen only for illustrative purposes.

The author would like to thank Aurel Sangenstedt for his comments on the draft of this chapter.

Expert evidence in price reviews and disputes

Christopher J Goncalves
Anthony J Melling
Berkeley Research Group LLC

1. Executive summary

In this chapter we focus on price reviews in long-term contracts ('LTCs') for the sale and purchase of natural gas and LNG.

The nature, scope and usage of price review ('PR') clauses in natural gas and LNG contracts has developed in tandem with natural gas market conditions in Europe and LNG market conditions worldwide. They are largely used to keep oil-indexed pricing mechanisms up to date with gas and LNG market conditions, as these industries have become increasingly independent of oil markets. As gas and LNG markets developed to become more interconnected, deeper, more liquid and more complex, the use of independent expert analysis has become increasingly critical for the conduct of PR disputes. Following the market liberalisation of the 1990s, European gas and LNG markets led the way in stimulating the use of PR processes with expert support. In recent years, Asian LNG markets have increasingly become more integrated both regionally and globally, and buyers and sellers are now making greater use of their own PR mechanisms, supported by industry experts.

1.1 Purpose

PR clauses and procedures serve to sustain the commercial viability of natural gas and LNG sale-purchase agreements over the long-term in the face of continuous market change. They provide a structure for Dispute Resolution ('DR') to maintain the viability of the commercial relationship, commercial bargain and allocation of price and volume risk between the parties.

1.2 History and outlook

In Europe in the early 1990s, gas market liberalisation drove the development and growth of gas hub trading. By 1997, the National Balancing Point ('NBP') was established giving the UK a single large virtual hub, and by 2008 the Dutch Title Transfer Facility ('TTF') became the main hub in continental Europe. Concurrently, a combination of high oil-indexed LTC prices and falling gas demand and hub prices created enormous price differentials between oil-indexed LTCs and short-term gas supplies sold by new wholesale competitors in

European markets. This led to significant losses from incumbent gas wholesalers, which triggered PRs to re-align their purchase prices with market conditions. Sellers argued that the traditional border netback prices should be revised, but buyers contended that a paradigm change to hub pricing should be implemented. As structured bi-lateral negotiations often failed, an increasing number of PRs were referred to international commercial arbitration. Experts were hired to produce expert reports on issues arising in these cases, including the relationship between oil-indexed LTC prices on the one hand and market prices and market value on the other.

In Europe, LNG PR clauses for LTCs are following the trends in European border supply contracts for pipeline gas, as LNG LTCs remain more heavily indexed to oil than to natural gas hub prices. In Northwest Europe ('NW Europe'),[1] LNG PR clauses will increasingly be used to bring oil price indexation slopes[2] into closer alignment with regional hub market conditions, and particularly the TTF. Outside NW Europe, oil price indexation remains the prevailing norm for European LNG LTCs and PRs have primarily served to bring LTC price indexation slopes into closer alignment with pipeline gas slopes and competitive gas market pricing.

In Asia, the legacy LNG markets have now seen several waves of PRs that have sustained the practice of oil price indexation, albeit at reduced oil indexation levels. As LNG markets become increasingly deep and liquid, and as the use of hub price indexation proliferates, Asian LNG PRs have become more complex and contentious. As a result, the use of external legal counsel, expert advisers, and international arbitration is increasing.

So far in Asia, the practice of PR is heavily linked to analysis of benchmark contracts for deliveries to national and regional markets considered relevant to the buyer and seller in their SPA PR clauses, but this approach confronts data transparency and quality issues that are even more onerous than pipeline border contract data issues in Europe. Benchmark contracts are also largely tied to oil, even though the levels of LNG competition are increasingly such that short-term market competition is an increasingly important factor in the analysis of market value of LNG. Unlike European natural gas markets, Asian LNG markets still lack liquidly traded price references for either LNG or natural gas. Nevertheless, competition is on the rise, and traders and policy makers are pushing various initiatives to create liquidly traded prices in the near future.

---

1 NW Europe includes the UK, Belgium, Luxembourg, Netherlands, Germany, Austria, Switzerland, Denmark and France.

2 In oil-indexation formulae, the variable element may be linked, for example, to the Brent Crude oil price in USD per barrel. The 'slope' or 'gradient' is a factor (expressed as a percentage) applied to the crude oil price to calculate the gas or LNG price in USD per MMBtu. For example, in an LNG SPA, at a crude oil price of USD 100 per barrel, and a contract price slope of 15%, the LNG price would be USD 15 per MMBtu. If the crude oil price fell to USD 50 per barrel, the gas price would fall to $7.50 per MMBtu.

1.3 Key practices and principles

SPAs are drafted to ensure that natural gas and LNG will be competitive with alternative fuels in the relevant market. PR provisions allow for regular revisions of the contract prices with the aim to maintain both the ability of the buyer to resell the gas, and a fair share of revenues to the seller throughout the contract term.

PR provisions define protocols with specific dates and time periods, including:

- discussion and negotiation windows.
- review periods for historical analysis of market change and prices.
- trigger date by which a PR can be initiated.
- application period for revised pricing (if any).

The first role of an expert in a PR may be to identify these key dates and define the relevant price review and application periods. Lack of clarity in PR clauses regarding the trigger mechanics, review period definition, price adjustment goals and parameters and related matters can produce substantial differences between the parties and their experts. Even when PR clauses are clear, market developments often present changes that test or challenge this clarity. As a result, it has become increasingly common in gas PR matters to engage external legal counsel and independent market experts to assist with the conduct of PR, and for the more intractable disputes with international arbitration proceedings. As LNG markets mature, this trend has also taken hold in the LNG industry, and is now on the rise in Asia.

1.4 Ingredients of success

In Europe, and increasingly in Asia, independent experts are used to provide market, commercial and economic analysis for a variety of roles: advising in informal price renegotiations, providing evidence for formal PRs, and providing analysis, reports and testimony as evidence in international arbitration.

The independent experts are most effective and efficient when engaged early in both the informal negotiation and formal review process. The delivery of a rigorous analysis of all available data, however imperfect, is a critical contributor to and can even be determinant of the overall success of client efforts.

This requires a robust market and commercial expertise in the relevant natural gas or LNG market, analytical rigour, intellectual integrity and independence, a sharp focus on the relevant facts at play in the market driving the PR and the ability to communicate the analysis clearly and concisely.

2. Purpose of price reviews

Natural gas is often sold under a LTC, also known as a Gas Sales Agreement

('GSA') or Gas Supply and Purchase Agreement ('GSPA'). Gas which has been liquefied may be sold under a Liquefied Natural Gas Sale and Purchase Agreement ('LNG SPA').

In broad terms, PR clauses and procedures, including international arbitration where needed, serve to sustain the commercial viability of natural gas and LNG sale-purchase agreements over the long-term in the face of market change. They are needed in large part because the traditional indexation of these contracts to established oil or fledgling natural gas hub index prices are not trusted to accurately represent market conditions and market value for the sale-purchase transactions over the full life of the contract.

PR clauses commonly serve two interrelated functions for long-term natural gas and LNG sale-purchase agreements. As summarised in the following sections of this chapter, these include:

- resetting contract pricing at market competitive levels, often by maintaining the commercial bargain between the buyer and seller over the long-term in the face of market change; and
- maintaining the allocation of price and volume risk between the parties over the long-term.

2.1 Maintenance of the commercial bargain between buyer and seller

The term of a LTC may range from 10 to 30 years, and may include extensions or renewal provisions. The duration of LTCs creates an intrinsic exposure of the contract pricing and commercial terms to market change and price evolution. PR and DR clauses recognise the need for LTC prices to be kept in line with market conditions over the contract term:

- The purpose of the PR mechanisms is to establish protocols to update the contract pricing in line with market conditions, often by seeking to maintain the purpose and intent of the original commercial bargain – typically with respect to the relationship between contract pricing, market conditions, and other non-price commercial terms in the LTC – made between the parties when they negotiated and signed the contract.
- DR clauses support the PR clauses by defining the protocols, procedures and mechanisms for DR if the parties cannot resolve a PR disagreement (or another disagreement) through bi-lateral negotiations.

Together, the PR and DR clauses provide the buyer and seller with a structured way to sustain their commercial bargain and maintain the commercial viability of their contract over the long term, in the face of potentially significant market and pricing changes that can produce disagreements between them as to the changing market value of the natural gas and LNG they purchase and sell under their LTC.

2.2 Maintaining the allocation of price and volume risk

A central component of the commercial bargain in many LTCs between buyers and sellers is the allocation of price and volume risk.

It is common for the seller to take the price risk (typically by subjecting its sales to the evolution of market pricing and value over the long-term), and the buyer to take the volume risk (typically by committing to take or pay most, or all of, the stated contract volumes over the long-term).

Price risk is typically allocated to the seller on the premise that the buyer requires competitive LTC purchase prices that it can profitably pass through to customers, over the long term. Historically, LTC prices were often linked to crude oil or oil products (using the principle that oil-products are the end-user's alternative fuel – known as the 'substitution principle' or 'netback pricing'). Under this practice, the seller assumed the risk of low oil prices and enjoyed the benefit of high oil prices.

Volume risk is typically allocated to the buyer. Sellers require their buyers to commit to take or pay for a specified volume of gas – often to ensure the reliable revenue stream needed to finance large scale investments in E&P and/or large scale pipeline transportation and/or LNG terminal infrastructure. Therefore, the buyer assumes the risk that gas market demand is not adequate to absorb all the contracted volumes, and is thus financially exposed to paying for gas it cannot sell.

However, this form of risk allocation has begun to break down. Over recent decades, the extent of natural gas and LNG industry infrastructure and market development has resulted in ever greater levels of market depth, liquidity, and competition. As markets become more competitive, the historical balance of price and volume risk has changed. Gas-on-gas competition has increasingly exposed buyers to price risk from competing gas supplies, and gas market liberalisation has provided buyers more options to mitigate volume risk, through improved access to infrastructure and/or traded market supplies.

In Europe, the implementation of the "Single Market" with gas-on-gas competition and market liberalisation resulted in LTC buyers assuming cross-commodity price risk. As traded gas market prices fell below those in their legacy oil-indexed LTCs for gas supplies, buyers were exposed to huge losses. Prolonged exposure to such losses posed an existential threat to some buyers, who had little option but to demand lower prices. PR mechanisms provided some buyers with a mechanism to achieve their aims.

In Asia, a similar scenario has begun to arise in the LNG industry, where buyers hold legacy oil-indexed LTCs at high slopes (some of which exceed 15%, approaching oil parity levels), but now have access to LTCs indexed to liquid natural gas hub prices in the US and increasingly abundant supplies of spot and short-term LNG traded at prices that increasingly resemble gas markets than oil pricing.

As illustrated in Figure 1, European oil-indexed LNG LTC prices substantially diverged from European spot prices during 2009–14, and then converged after 2014. This was due to the oil price collapse in late 2014 and 2015, coupled with declining oil indexation slopes in new LTCs executed during 2013–14. In Asia, oil-indexed LNG LTC prices diverged from JKM[3] DES[4] price assessments for spot cargoes during 2009–11 and then converged after 2011. This was largely due to the LNG demand shock caused by the Fukushima nuclear incident and Japanese nuclear shutdowns, which drove up JKM prices to lure LNG cargoes to Japan. These prices reached parity with the legacy high sloped oil-indexed LTCs in Asia. With increasingly ample supply after 2014, JKM has fallen back below LTC prices, but by a smaller margin than before Fukushima.

Figure 1. Oil-indexed LNG LTC DES prices and assessed DES spot prices[5]

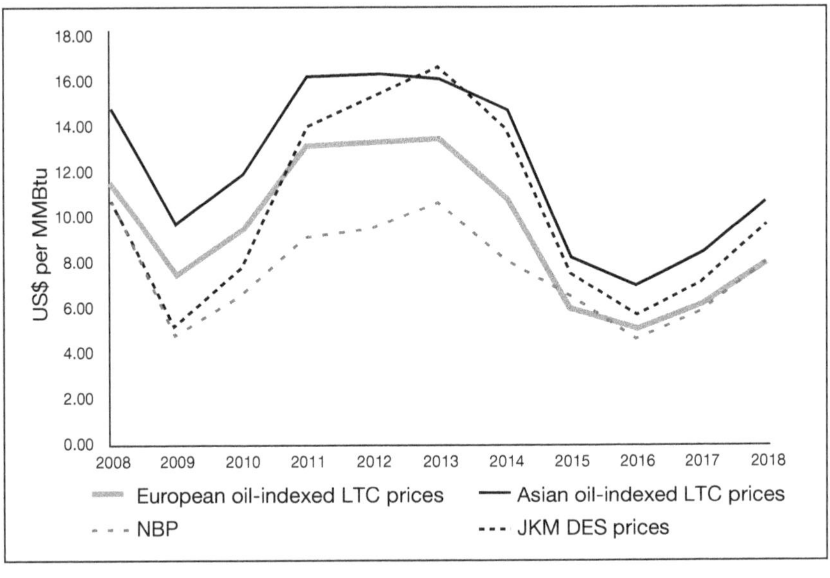

3. Historical evolution of PR clauses

The use, nature, and complexity of PR clauses has evolved in tandem with gas and LNG markets, policy, and regulation primarily in Europe and Asia. As gas and LNG markets have become deeper, more liquid, more complex and

---

3 Japan Korea Marker Price Assessment published by Platts. JKM is used as a price reference for spot cargo trades in Asia.

4 Delivered ex-ship. A shipping term meaning that the seller is required to deliver LNG to a specified destination.

5 BRG Energy analysis based on market intelligence and third-party resources including ICIS Heren, Poten, Global LNG Info, Wood Mackenzie, Bloomberg and Platts. LTC DES Prices are calculated based on two steps: first, we calculated the delivery volume weighted average slopes and constants of LTCs (of 10 years or longer) for each year from 2008 to 2018; secondly, we calculated the LTC DES prices based on historical Brent prices, and weighted average slopes and constants from step one.

increasingly independent of oil markets, PR clauses have been developed and refined to address and resolve the increased cross-commodity price risk exposure presented by sustained use of oil price indexation for LTCs.

3.1 For European pipeline gas

Historically, the European gas industry developed the principle of netback pricing to national borders. To capture market share, gas companies had to incentivise customers to invest in natural gas equipment through competitive pricing against established fuels, mostly light and heavy fuel oils. Netback pricing calculations began with the 'burner-tip' market value of natural gas in inter-fuel competition, based on the substitution principle. Deductions were made for the costs of transport services, overheads and profit margin to arrive at the 'netback value' at the point of sale (typically a national border). Each sector of the buyer's market was examined separately, and a calculation made of the weighted average netback value at the border.

This formulaic approach to the calculation of the initial market value provided a solid starting point for the mechanism of periodic PRs. Parties could replicate the original netback calculations, often with the assistance of experts who specialised in the dynamics of end-user oil product prices and calculation of the costs in-between the burner-tip and the point of sale at the national border. In support of the price revisions, prices were frequently benchmarked against national statistical data, or EU sources such as Eurostat end-user prices. PR clauses provided for this periodic updating of the netback value at the border, and the calculations were often a critical element of the expert's role.

However, there were many cases of problems with the information available for relevant markets:

- In some countries, the data did not exist – few countries published a full range of benchmarking information from end-user prices, midstream costs and border prices.
- In other countries, the available data was considered unreliable – gaps were known to exist in some national statistics because some key market participants did not respond to survey requests.
- Benchmark contract information was often lacking in detail – although price data might be available, other terms were missing (such as contract duration, load factor, and indexation formulae).
- Some data was not considered arms-length because it included elements such as transportation costs that were paid to associated or subsidiary companies.

By rigorous analysis and interpretation of the full range of available statistical evidence, gas industry experts apply their skills to overcoming the challenges presented by incomplete or flawed data. In some cases, PR analysis required the LTC price to be benchmarked against other published border prices from other countries. For example, the German Border Price ('GBP') could also be cited in PR analysis of surrounding countries. In other cases, independent databases of contract price information were used.

In Europe, the netback pricing paradigm was eventually rendered obsolete by the emergence of hub[6] price mechanisms. The PR backdrop began to change with natural gas market liberalisation from the late 1980s, leading to the development of gas-on-gas competition in the early 1990s, the exploitation of new supplies, and the emergence of gas trading hubs in 1995–96. In market conditions of gas oversupply, hub prices could settle at levels much below netback prices, creating an existential threat to buyers heavily exposed to netback prices that were far above competitive hub prices.

As the liquidity of continental spot markets increased in 2008–09, a combination of high oil-indexed LTC prices and falling gas demand and hub prices created enormous price differentials between oil-indexed LTCs and short-term hub prices (see Figure 2).

Figure 2: UK hub prices (NBP) v oil-indexed prices (represented by GBP)

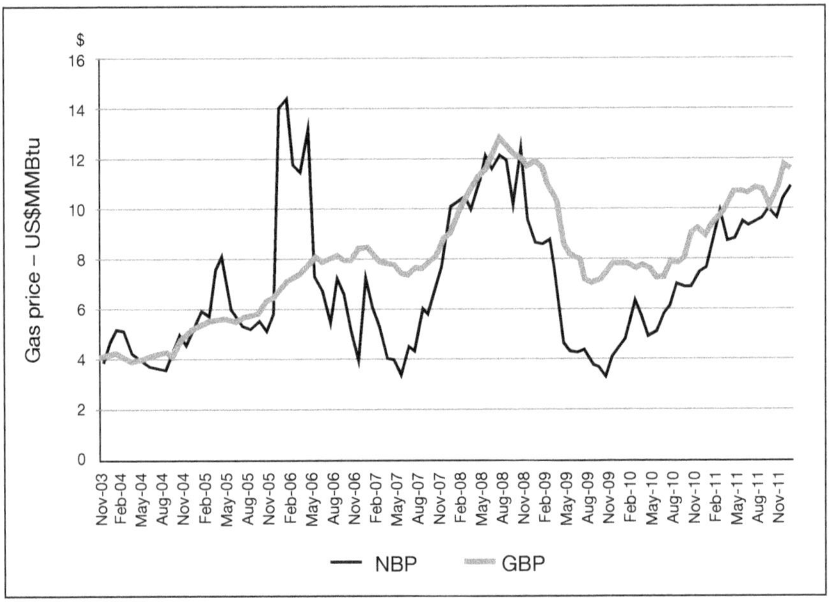

In this changing environment, buyers with netback pricing mechanisms increasingly demanded structural price indexation changes to prices and PR

6 Hubs can be either a meeting place of multiple pipelines and/or terminals (ie, a physical hub) or a demand area with many suppliers and customers (ie, a virtual hub).

clauses. This change was straightforward to implement in new contracts, but under existing contracts prospective price adjustments needed the analytic support of market and commercial experts and contract interpretation by legal counsel.

Today in NW Europe, the liquidity of spot markets has facilitated a transition from oil-indexed to hub-based pricing. Extensive use of PR procedures and PR arbitrations has wrought a near complete paradigm shift from oil-indexed pricing to hub pricing. Some legacy contracts with oil-indexation still exist, but the price levels have generally been aligned to near spot market levels.[7] With LTC prices linked to hub prices, the economic utility of adversarial PRs has decreased.

In other areas of Europe, notably Iberia and Eastern Europe, the transition towards hub pricing remains incomplete, and in Asia the transition remains in the early stages. The spread of hub indexation has been impeded by pipeline connectivity and the relative lack of access to competing supplies. As new EU interconnection projects are implemented, and new LNG import terminals built, access to the liquid hubs of NW Europe is improved, leading to dramatic declines in the price differentials between those markets.

In the markets where connectivity to liquid hubs is poorest, there may be no clear, single indicator of market value. Therefore, carefully drafted PR clauses and procedures remain critical to managing cross-commodity risk, and future-proofing LTCs.

3.2 For European LNG

Before the 2000s, European LNG LTC pricing was linked to either crude oil or crude oil product prices. During the 2000s, PR clauses increasingly began to reflect market liberalisation, gas market pricing and hub indexation — first in Europe, and more recently in Asia (see Table 1 on the next page).

As European pipeline gas pricing conditions changed, LNG SPAs have followed. In NW Europe, LNG prices have been gradually rebased to levels that are competitive with pipeline gas and hub markets through PR or arbitration, and some existing oil-indexed mechanisms were replaced with gas hub indexation, particularly to NBP and TTF, through arbitration. New contracts are hub-indexed.

Beyond the UK and NW Europe, the absence of liquid gas hubs makes it challenging for buyers and their legal counsel to argue for replacing oil price indexation with hub market indexation. In the Iberian Peninsula, PR clauses have successfully been used to adjust LNG SPA prices, but oil-indexation

7 Luca Franza, "Long Term Gas Import Contracts in Europe – The Evolution in Pricing Mechanisms" Clingendael International Energy Program report (2014). Page 15 notes that the first report of hub-indexation by Gazprom arose in 2010, when it was reported that 15% of gas volumes sold to E.ON would be tied to European spot prices. It also notes Gazprom's use, in other cases, of 'Corridor Pricing', a virtual hub price where the core formula is oil indexed, but the price is maintained within a narrow band by a top stop, and a bottom-stop formula that are linked to gas spot prices. In effect, this is a form of hub pricing disguised as oil-indexed pricing.

Table 1. Hub indexation share of total LNG LTC volumes by region[8]

Hub indexation share of total LTC volumes	2008	2018	2025
Europe	7%	19%	55%
Asia	0%	5%	22%

remains alongside some SPAs that use hub price indexation. In Eastern Europe, as the new LNG contracts begin to come up for PR it is likely that increased LNG market competition, LNG competition with monopoly pipeline gas and increased downstream gas market competition will play a greater role in the adjustment of oil indexation slopes in PRs and disputes, but it will be some time before downstream regional gas hub market depth and liquidity are sufficient to support the replacement of oil indexation with hub pricing.

Contracts for the purchase of LNG from North America which have been signed since 2011 are beginning to impact European LNG PRs. Prices in these contracts have a variable element that is linked to the Nymex HH spot gas market. In countries where deep and liquid gas hubs exist, LNG deliveries are often re-sold at local hub prices so the HH pricing of LNG supplies has no direct price effect, but rather an indirect effect of the additional volumes on the supply-demand balance of local hub trade and pricing. However, in other parts of Europe where liquid hubs are not yet established, the arrival of US LNG with HH pricing has a more direct and significant price impact that buyers, sellers, and their experts increasingly need to factor into PR analysis as a source of material market change.

3.3 For Asian LNG

The practice of indexing LNG prices to competing oil products developed in both Europe and Asia. In both regions, market growth relied heavily on increasing international trade in natural gas and LNG. However, there was one fundamental difference between these regional markets. In Europe the supplies were strongly connected to the markets, and markets connected to each other, mostly by pipelines. But, for geographical reasons, in Asia the supplies were connected to markets mostly by LNG vessels, with few international pipelines. As a result, the European LNG business remained relatively small and subordinate to pipeline commercial practices; whereas, the LNG business in Asia thrived as the primary form of international gas trade.

8 BRG Energy analysis based on market intelligence and third-party resources including ICIS Heren, Wood Mackenzie and Global LNG Info.

The evolution of LNG LTC annual contract quantities ('ACQs') in Europe and Asia is summarised in Table 2:

Table 2. LNG LTC ACQs by region[9]

LTC Capacity (Bcma)	2008	2018	2025
Europe	36	80	87
Asia	103	165	249

The regional differences between Europe and Asia, and the geographical distance and shipping costs between the LNG markets, limited the regional cross-fertilisation of LNG commercial structures and fostered the development of differentiated pricing mechanisms and PR clauses in Europe and Asia.

In Europe, buyers often demanded gas price discounts that were sufficient to incentivise their downstream customers to switch from other fuels, and deductions to reflect the costs of gas delivery to the burner tip.

By contrast, Asian markets developed a reputation as 'premium' markets for LNG sales. The development of LNG sales in Asia in the 1960s and 1970s was heavily based on sales into the Japanese power generation sector, where the government was seeking to reduce dependence on oil-fired generation. Power generators were willing to switch to natural gas, which could be burned more cleanly and efficiently than oil. These customers typically appreciated even a small discount to crude oil prices. To provide some protection from oil price spikes the Japanese buyers negotiated S-curve contracts which reduced the impact of high oil prices, in return for providing the seller with downside protection in a low oil price environment.

As other Asian buyers such as Korea and Taiwan emerged, the 'big three' were collectively referred to as the 'JKT' market. New players generally adopted commercial structures that were similar to the Japanese model, especially in the price and PR clauses. Japanese LNG prices were nearly always linked to Japanese customs prices for imported crude oil but other countries, such as Korea, had the option to link to their own oil import prices.

Asian LNG markets involved LTCs from a relatively small number of sellers to a small number of buyers. Historically they were less liberalised, competitive and complex. JKT markets maintained oil price indexation through several waves of PRs for legacy contracts. These markets had no supply alternatives from domestic gas production or pipeline imports. As a result, they have no

9 *ibid.*

domestic gas market conditions or gas market pricing to reference for purposes of PR and price adjustment.

Many JKT LTCs have PR clauses, but some do not. Those that do have PR clauses typically reference benchmark LNG import contracts into the relevant country or region during a historical review period as the basis for measurement of price adjustments. Parties compared their contract prices to pricing estimates of the confidential pricing for comparable benchmark contracts. These estimates were typically procured from third-party energy data providers. As a result, LNG contract benchmarking was relatively straightforward. The contract parties typically performed this analysis internally without use of independent market and commercial experts, or legal counsel. PRs were largely settled by the parties through the review process without resorting to international arbitration or other forms of DR.

Although this approach was lower cost, the results were not economically efficient for Asian LNG buyers. They paid prices that were often higher than European price slopes by between 3% and 6% of Brent oil prices. Those differentials became very painful when oil prices reached sustained levels above $80, and sometimes well over $100 per barrel during the time period from 2010 to late 2014. At oil prices of $80 per barrel, this meant that Asian LNG buyers paid prices approximately $2.4 to $4.8 per MMBtu more than European buyers.

In periods when market conditions shifted in favour of buyers, LNG spot market prices (as assessed by Platts in the Japan Korea Marker or 'JKM' price assessment) declined and the negotiated oil price slopes of new benchmark contracts tended to decline, and vice versa. With the embedded lag of the retrospective PR process, the changing conditions for new benchmark contracts tended to get picked up in PRs. In cases where the parties could not agree on oil slope adjustments, but did not wish to resort to arbitration, the solution was sometimes to reduce the contract ACQ and/or increase the contract flexibility provisions. This allowed the buyers to make greater use of lower priced supplies from new contracts and/or spot markets while sellers maintained the existing oil indexation slope.

Since the oil-price collapse of late 2014 and early 2015, and the ongoing arrival in East Asia of sequential waves of vast new LNG supplies from Australia, Papua New Guinea ('PNG'), Russia and the US, supply conditions have again shifted in favour of buyers. On the demand side, the emergence of price sensitive customers such as China and India, and the growth of Asian LNG spot markets has brought new demand, but also opened new debate about premium prices in Asia. The result of ample supply and elevated buyer price sensitivity is that the 'Asian premium' in LNG spot markets has been almost eliminated as JKM prices have plummeted to levels that are much closer to European gas hub prices. Further, the indexation slopes for new LNG benchmark contracts have fallen by several percentage points, the duration of new LTCs has decreased and the flexibility of new LTCs has also increased.

As shown in Figure 3, LTC and spot prices have broadly declined and converged over the last decade. During 2008–09, oil slopes for new LTCs concluded for delivery to Europe and Asia averaged 0.13 and 0.15, respectively. However, LTC oil slopes declined to 0.11 and 0.12 for Europe and Asia, reaching similar levels to the imputed oil slopes for spot market prices.

Figure 3. Historical spot prices and imputed slopes compared to average slopes in new LTCs[10]

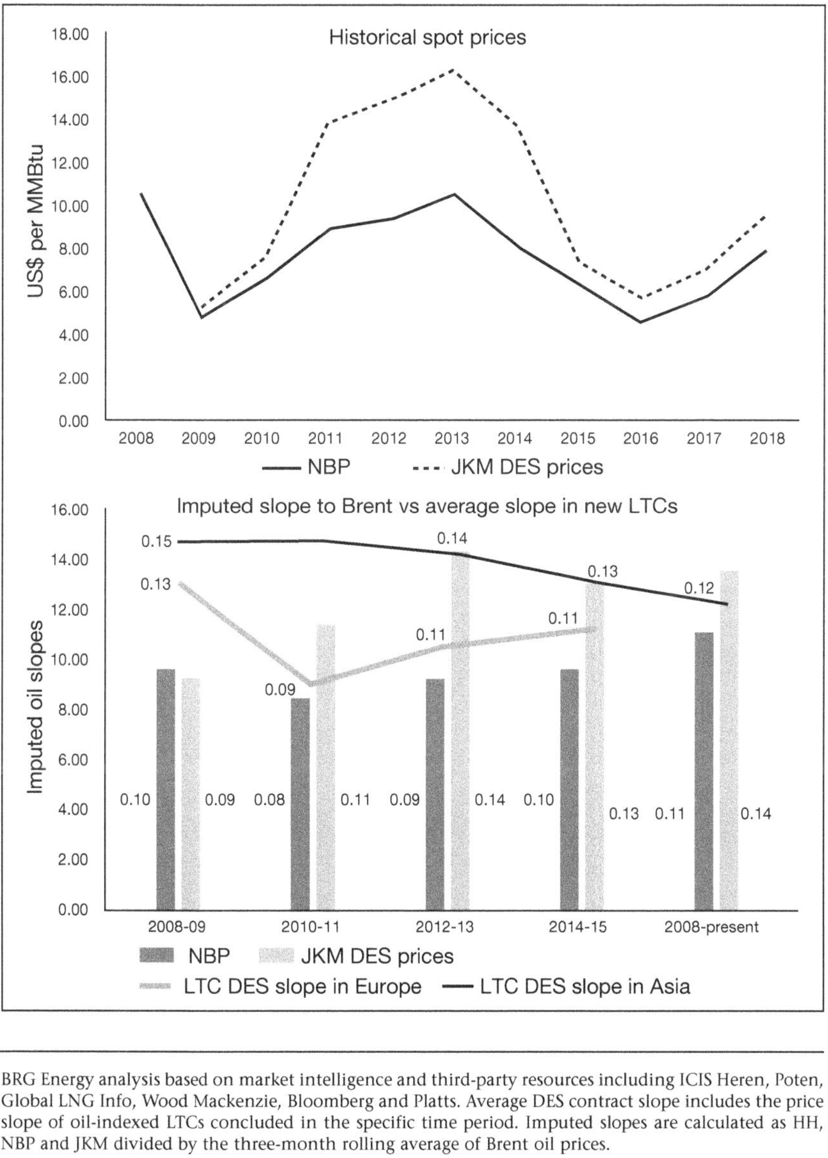

10 BRG Energy analysis based on market intelligence and third-party resources including ICIS Heren, Poten, Global LNG Info, Wood Mackenzie, Bloomberg and Platts. Average DES contract slope includes the price slope of oil-indexed LTCs concluded in the specific time period. Imputed slopes are calculated as HH, NBP and JKM divided by the three-month rolling average of Brent oil prices.

Also, a substantial volume of new Asian LNG contracts have been indexed to the US Henry Hub[11] ('HH') price, introducing extra-regional gas hub indexation into the fleet of benchmark contracts. Churn rate is a good measure of a trading hub's liquidity and commercial success. In 2018, the churn rate of HH reached 53.9, which is much higher than the churn rates of NBP and TTF (see Table 3). This indicates that HH has become very liquid.

Table 3. 2018 Total traded volumes and churn rates[12, 13]

Hub	Country	2018 Traded volumes (Bcm)	2018 Churn
HH	US	46308	53.9
NBP	UK	1359*	16.9
TTF	FR-DE-AT-CZ-Benelux	2540	16.7
JKM	JKT and China	53	0.2

* Based on Patrick Heather's figure of 15105 MWh

These market changes and increased market complexity have led buyers and sellers to regularly engage independent experts and legal counsel for their PRs, and to more readily consider the use of international arbitration to resolve their more challenging disputes.

These market conditions have changed the Asian LNG contract markets, and contract PR conditions, in several ways:

- First, the substantial reduction of spot prices and oil indexation slopes for new LTCs has been substantial, appears to be sustainable over the long-term, and thus has raised the economic stakes for PR substantially, a lot of value and money are now on the table.
- Secondly, the indexation provisions for new benchmark LTCs have become more complex due to the advent of HH pricing alongside oil indexation presenting a challenging new cross-commodity problem for buyers and sellers to grapple with in PRs.
- Thirdly, the trend toward shorter and more flexible contracts, and the

11 Henry Hub is the world's deepest and most liquidly traded natural gas hub.

12 "Quarterly Gas Review, Analysis of Prices and Recent Events", The Oxford Institute for Energy Studies (March 2019) p10: www.oxfordenergy.org/wpcms/wp-content/uploads/2019/03/Quarterly-Gas-Review-5.pdf.

13 Patrick Heather, "European and Global Hub Markets: What Does the Future Hold for Hubs in the Spot-priced Gas Market?" Gas and LNG Supply Contracts Forum (19 June 2019).

greater use of lean LNG from the US in markets typically dominated by rich LNG, has presented important non-price commercial features of benchmark LTCs for the parties to address in PRs.

These recent market and commercial trends in the Asian regional LNG trade have fundamentally changed the LTC PR landscape. Whereas in the past PRs could be resolved between the parties through the direct negotiation of modest slope, duration, and/or flexibility adjustments; the gap between buyer and seller views regarding pricing and flexibility is now quite substantial, harder and more complex to resolve, and in some cases intractable.

In this environment, it is now common for LNG buyers and sellers to engage the assistance of external legal counsel and independent experts before and during the PR process to assist with the review and provide advice regarding the opportunities and risks associated with international arbitration. Although the preferred outcome remains to settle PRs through bilateral negotiations, this has proven very difficult. Therefore, a growing number of buyers and sellers are now engaged in detailed evaluation of prospects for taking their PR disputes into international arbitration.[14]

Looking forward in Asia, LNG PRs will continue to become more complex, contentious and rigorous, such that the use of external legal counsel and advisers will increase. It is also likely that the use of international arbitration will increase as buyers become more familiar with this tool as a useful resource for solving complex problems that the parties' internal negotiators cannot resolve. Nevertheless, arbitration may not become as widely used in Asian LNG as it has been in European pipeline gas and LNG, primarily because in Asia many contracts will expire over the coming several years, giving buyers a commercial opportunity to rebase oil slopes and/or change price indexation without engaging in legal disputes.

4. The role of the expert in the PR process

After several decades of development, today's natural gas and LNG industries include multiple stages of commercial exchange and transportation from exploration to end-use, sometimes referred to as the 'value chain'. These industries are complex, and geographically and structurally diverse.

As a result, expert evidence has become a central component of natural gas and LNG PRs in Europe. Most European PRs and PR arbitrations are conducted around a core of expert analysis and evidence, and where needed direct oral testimony in arbitration proceedings.

---

14 The first reported case of an Asian PR dispute being taken to international arbitration was reported by Reuters on 12 February 2018: "South Korea's KOGAS in LNG arbitration with Australia's North West Shelf Gas" (www.reuters.com/article/us-southkorea-kogas-north-west-shelf/south-koreas-kogas-in-lng-arbitration-with-australias-north-west-shelf-gas-idUSKBN1FW0DU).

In our experience, the PR process is best served when independent experts become engaged early in both the informal negotiation and formal review process. This enables the expert to better understand and advise on the unique opportunities and requirements of informal negotiation, formal PR and international arbitration. He/she can then be positioned to advise on the economic tradeoffs and risks attendant with negotiated settlements, on the one hand, and formal reviews and disputes on the other. Across all the potential expert roles, rigorous, objective analysis is a critical contributor to the overall success of PR efforts. Best practices are built upon several critical qualities:

- demonstrated market and commercial expertise in the relevant natural gas or LNG market;
- analytic rigour and thoroughness of checking and cross-checking data, facts, and analytic assumptions and results;
- intellectual integrity and consistency across engagements and engagement tasks;
- intellectual independence and objectivity, including the willingness and ability to challenge or differ from client assumptions when needed;
- a sharp focus on relevant facts and information over marginal or unimportant detail; and
- the ability to communicate analysis and conclusions accurately, clearly and concisely across graphic, written and oral submissions.

These qualities lead to analytic results, reports, and/or testimony that have the rigour, credibility and independence needed to persuade counterparties and/or convince arbitration tribunals of recommended pricing solutions in complex circumstances.

The expert does not provide a legal interpretation of the PR or DR clauses in LTCs. However, the experienced expert's understanding of industry mechanics and practices often provides useful support to assist the lawyer and/or arbitrators in their interpretation of these clauses.

4.1 Mechanics

It is common for PR mechanisms to be constructed with multiple stages. First, notification of intention to open a PR must be given. The party initiating a PR must have a contractually valid reason for the review, such as a material change in market conditions that has a substantial and sustained impact on prices. Following notification, the parties are obliged to negotiate in good faith and, if agreement is not reached by a specified point in time (typically six months), the dispute may progress to either expert determination or arbitration (depending on the DR clause). Occasionally, where agreement looks both imminent and likely, the parties may jointly agree to extend the negotiation period. Key issues addressed by the expert include:

(a) ***Review period conventions***

The review period is critical to most PR disputes. DR clauses often allow PRs to start once every three years, or every five years for some LNG contracts. Additional PRs may also be possible through so-called 'joker clauses' that allow for one or two unscheduled PRs to be initiated in exceptional circumstances. The review period for market changes is generally the three-year period preceding each PR start date.

(b) ***Trigger mechanics and standards***

The initiation or 'trigger' of a PR is typically at the option of either party, and requires written notification by the party initiating the PR before a stipulated 'trigger date'. If a PR is not activated, then it is typically inferred that market changes were insufficient for either party to trigger a PR. Market changes that are deemed to be sufficiently material and/or sustainable, rather than brief and fleeting, are typically considered valid 'triggers' for PR.

(c) ***Adjustment periods and practices: rebase slope or index change?***

The powers available to arbitrators may be limited under the terms of the PR clause to exclude specific outcomes, such as the possibility of an index change. In such cases the only option available to the arbitrator may be to adjust the indexation slope and/or fixed factors[15] under the price indexation formula.

The criteria for triggering a PR may be different from the criteria for price adjustment and claiming damages. Even where a trigger event has been agreed or awarded, the case for price revision and/or damages must start from first principles according to its own separate PR clauses and protocols. Further, while the market change for triggering PR should be material, some LTCs impose collars around the amount of price adjustment allowed as a result of PR.

Lack of clarity in PR clauses as the trigger mechanics, review period definition, price adjustment goals and parameters and related matters can produce substantial differences between the parties. Even when PR clauses are clear, market developments often present changes that test or challenge this clarity.

4.2 Key issues and principles

While natural gas and LNG contracts, and their PR procedures, can vary widely across regions and countries, there are some broader PR principles and practices that apply:

(a) ***Comparing market change to LTC pricing***

First, the fundamental purpose of PR for both natural gas and LNG is to measure

15 Both oil-indexed and hub-indexed LTC price formulae often contain two distinct price elements: the commodity-indexed component (generally oil or gas indexed), and a fixed component (often reflecting a seller's fixed cost such as transportation fees or LNG tolling charges).

market change during a historical review period (see above) and compare any such change to the contract pricing over the same period. Depending on the expert's instructions, the expert may be required to analyse and explain for a given review period the dynamics of markets, supply/demand fundamentals, price trends, transportation costs and the status and impact of liberalisation and competition, among many other potential issues. If market conditions have changed and contract price levels and/or price indexation practices have either not kept up with or have moved ahead of the market, then a price adjustment is merited to bring the contract price levels and/or indexation back into alignment with market conditions for the next pricing period (see above).

Of course, the definition of the subject market can differ between natural gas and LNG contracts, and market definition is sometimes a point of contention between the parties (see below).

(b) *Data sources and data quality*

As noted above, the practice of price review has changed and improved in Europe as the depth of hub trading and quality and reliability of market information has improved. In the more developed markets, complex analysis and calculations of netback pricing and border contract price benchmarking have given way to comparison of contract pricing to hub market indicators of market value.

This has not yet occurred in Asia and data sources and quality remain a material issue. As noted, in the absence of liquidly traded competitive downstream markets for natural gas, the practice of Asian LNG price reviews is most commonly linked to the analysis of benchmark LNG contract pricing. This confidential contract pricing data is available from a few third-party vendors, who compile it based on discussions with the parties and educated estimates or inferences. The accuracy of this data is a critical issue.

In our experience of several Asian LNG PRs for leading buyers and sellers (who have multiple contracts), we have compared client contract data to third-party database information on several occasions. We have found consistent material discrepancies between the pricing reported by data vendors and client contracts.

For example, in recent Asian LNG PRs, we have found cases where the vendor database information correctly reports oil indexation slopes, and many others where the vendor information was overstated by over 0.5% of slope. For pricing constants, we have seen a wide range of errors ranging from \$0.40 per MMBtu above or below the actual contract pricing constant. In one case, we noted that third-party data sources over-stated contract prices by more than \$25 million per year.[16] Comparing all contracts we have reviewed to third-party data, the net overestimation would be between \$12 to \$14 million.

16 Assuming an average oil price of \$75/bbl, and assuming a conversion of 46.405 MMBtu per tonne.

There is no easy solution to the errors embedded in the third-party assessments of confidential contract pricing data used for price benchmarking in Asian LNG PRs. Cross-checking and/or averaging available data sources is helpful because vendor data errors are not directionally consistent. In some cases, countervailing vendor data errors tend to offset each other, where one is higher and the other lower than the actual contractual price. In other cases, one vendor's information may be factual and accurate; whereas another's may be estimated and inaccurate.

Assuming that Asian LNG and natural gas trading hubs develop comparable depth and liquidity to the levels achieved in Europe, liquidly traded Asian LNG and natural gas markets will eventually replace vendor data benchmarking for LNG PRs, as previously occurred in Europe.

(c) *Market definition and measurement*

Secondly, market definition is critical. Most PR provisions aim to ensure that natural gas and LNG will be competitive in the relevant end-user market.

The substitution principle (see section 3.1), originally used to establish oil-indexed natural gas and LNG pricing in both Europe and Asia, is based on the notion that natural gas is purchased to displace oil and petroleum products and should be priced at a discount to those products.

As energy and natural gas markets have developed away from reliance on petroleum products and toward greater usage of natural gas and renewable energy, the substitution principle rarely applies solely to oil. Oil-indexed LTCs now compete with one or more of:

- other natural gas or LNG supplies sold on the basis of spot or hub-traded prices;
- coal supplies for power generation;, and/or
- new renewable energy resources.

This makes it more complex and contentious to determine whether and how a natural gas or LNG price may be needed to bring contract pricing back into alignment with market conditions.

Unless the contract PR provision is very clear and prescriptive, buyers and sellers frequently disagree about whether the comparative measurement of market change should concentrate on the downstream market into which the LNG buyers utilise or re-sell natural gas, the border market for imported pipeline gas or beach market for imported LNG supplies, or even the global LNG market. Experts are frequently tasked to provide clarity on this issue.

Nevertheless, it is broadly true that many natural gas LTCs, and some LNG SPAs, reference the buyer's downstream market. Other LTCs refer to national or regional borders, or DES markets for LNG under benchmark LTCs and/or spot market conditions, especially in Asia.

For Asian LNG LTC benchmarking, one critical difference in reference market definition can be whether the PR clause calls for comparison to LTCs that were 'executed' or 'operational' during the recent years during the review period. Evaluating recent contracts based on the date in which they were executed provides for analysis of the most recent prices of new contracts; whereas analysis of recently operational contracts includes a larger list of legacy contracts that are still in effect. The former approach tends to capture the most recent trends in the market, and the latter approach tends to capture a more stable long-term trajectory of market value. In the recent environment of ample supply, the difference between PR clause references to 'executed' and 'operational' contracts can produce very different results. PR clause references to 'executed' contracts tend to produce lower prices than PR references to 'operational' contracts, but this dynamic could reverse in a market with tightened supply, rising prices, and rising slopes for new LTCs. Where PR clauses are unclear on such issues, expert analysis is often used to determine which approach is most applicable.

(d) Buyers need to economically market the gas

Thirdly, PR clauses may or may not include a 'Hardship Clause' which provides that the buyer should be able to economically re-sell the gas in a specified market. However, it is broadly true that LTCs for both natural gas and LNG typically reference or imply the buyer's need and right to be able to economically or competitively market the gas or LNG purchased and sold under the contract. Sellers do not generally oppose this principle because sustained above-market contract prices would make the contract unviable or unsustainable.

However, a seller and buyer may disagree over the interpretation of hardship clauses, how the price competition should be measured and whether it implies or guarantees the buyer's right to earn a margin and/or the size of that margin. They may also disagree about the relevant duration of hardship. Temporary hardship is a market reality faced by most businesses, but prolonged hardship can pose an existential risk. Should the hardship be measured over the RP, or is a trend of declining profitability better evidence? In the event of disagreement, parties may instruct their experts to analyse the extent of the buyer's ability to earn margins, typically by the review of netback calculations from the specified market to the point of sale under the LTC.

(e) Materiality and sustainability of market change

Finally, there are two other principles regarding the materiality and sustainability of market change (as noted in section 4.1) that are not as broadly referenced in LTCs or evaluated by the parties to a PR, but are often considered by arbitration tribunals. Once it is determined that market change has occurred and that the relevant market pricing conditions have diverged from the contract pricing, there are important materiality and sustainability tests that some arbitrators often evaluate alongside potential contract price adjustments.

If the divergence between the contract price and relevant market is quite small, it may not result in a price adjustment. If the measurement methodologies were strongly debated and/or the calculations were very complex, a tribunal has the option to decide that the change is not material rather than make a very small and potentially inaccurate change.

Further, even if the market diversion from the contract pricing was substantial for a small or large period during the full historical review period, but then either realigned with contract prices at the end of the review period and/or just after the review period, then the diversion may not be considered to have been sustainable. Some parties to PR and/or their arbitration panel may not wish to implement a price adjustment if the market diversion from the contract pricing was clearly temporary and has already disappeared or resolved.

The adjustment phase in gas price reviews under long-term gas supply contracts

Michelle Glassman Bock
Steven Finizio
Wilmer Cutler Pickering Hale and Dorr LLP

1. Introduction

While the precise steps and procedure will vary according to the terms of the parties' contract, gas price reviews under long-term gas supply contracts are often described as having two phases: (1) trigger and (2) adjustment. This chapter focuses on the adjustment phase.

During the adjustment phase of a gas price review, the parties address what the appropriate adjustment to the contract price should be. That issue needs to be decided only if the parties agree or an arbitral tribunal decides that the prerequisites for a price revision have been met.[1] Moreover, what adjustment is appropriate will often depend on what is found to have triggered the need for a revision. However, because the contractual requirements for a price review can vary significantly, that means the adjustment phase can likewise vary significantly.

The sections below discuss different contractual approaches to price adjustments in long-term gas supply contracts with a particular focus on import contracts. Section 2 describes the approaches taken in two publicly available price review clauses; section 3 analyses the relationship between trigger and adjustment; section 4 looks at the scope of an adjustment; section 5 considers the application of any relevant benchmarks in determining the adjustment; section 6 assesses the impact, if any, of an 'in any case' clause; and section 7 discusses certain procedural issues that can arise with regard to the adjustment phase of a price review arbitration.

2. Different contractual approaches to the adjustment phase

The starting point in determining an appropriate adjustment is what the price review clause says about the adjustment and, in particular: (a) whether it

1 While disputes about gas price reviews can also be decided by courts or by expert determination depending on what the parties agree, most long-term import contracts refer such disputes to international arbitration.

provides guidance on how an adjustment should be determined, and (b) what, if any, benchmarks (ie, price data other than the contract price) should be considered.

Given the confidential nature of long-term gas supply contracts, there are very few publicly available price review clauses. However, two such clauses, set out below, help illustrate some of the different approaches parties take to the adjustment phase of a price review.

One of the best-known price review clauses is from *Atlantic LNG Company of Trinidad and Tobago v Gas Natural Aprovisionamientos, SDG, SA*. The clause provides:

(a) If at any time either Party considers that economic circumstances in Spain beyond the control of the Parties, while exercising due diligence, have substantially changed as compared to what it reasonably expected when entering into this Contract or, after the first Contract Price revision under this Article 8.5, at the time of the latest Contract Price revision under this Article 8.5, and the Contract Price resulting from application of the formula set forth in Article 8.1 does not reflect the value of Natural Gas in the Buyer's end user market, then such Party may, by notifying the other Party in writing and giving with such notice information supporting its belief, request that the Parties should forthwith enter into negotiations to determine whether or not such changed circumstances exist and justify a revision of the Contract Price provisions and, if so, to seek agreement on a fair and equitable revision of the above-mentioned Contract Price provisions in accordance with the remaining provisions of this Article 8.5.

(b) In reviewing the Contract Price in accordance with a request pursuant to sub-Article 8.5(a) above the Parties shall take into account levels and trends in price of supplies of LNG and Natural Gas [in redacted market] such supplies being sold under commercial contracts currently in force on arm's length terms, and having due regard to all characteristics of such supplies (including, but not limited to quality, quantity, interruptability, flexibility of deliveries and term of supply).

(c) The Contract Price as revised in accordance with this Article, shall in any event, allow the Buyer to market the LNG supplied hereunder in competition with all competing sources or forms of energy (including, but not limited to, natural gas, fuel oil, gasoil, coal, LPG, district heating and electricity) … in the market of the Buyer at the point of consumption …, taking into account, inter alia, all appropriate operations, services and risks which are usual in the Natural Gas industry from the points of import for handling and marketing the Natural Gas in all market segments when due regard is given to all characteristics of the LNG supplied under this agreement (including, but not limited to, delivery point, quality, continuity, flexibility of deliveries and term of supply) … and on the basis that sound marketing practices and efficient operations on the part of

the Buyer are assumed and such Contract Price shall allow the Buyer to achieve a reasonable rate of return on the LNG delivered hereunder.[2]

Another frequently cited price review clause is from the Final Award in ICC Case No 9812:[3]

(a) If the economic circumstances in the country of the Buyer which are beyond the control of the Parties should change significantly compared to what is reflected in the prevailing price provisions ..., then each Party shall be entitled to an adjustment of the price provisions ..., reflecting such changes, in particular the value of Natural Gas in the end user market of the buyer as such value can be obtained by a prudent and efficient gas company.

In any case the price provisions hereunder shall allow the Buyer to economically market the Processed gas delivered hereunder in the market of the Buyer in competition with all competing sources of energy in the end user market always assuming sound marketing practices and efficient operations on the part of the Buyer.

Some of the differences between these clauses, and some common issues that arise at the adjustment phase of a gas price review, are discussed below.

3. Relationship between trigger and adjustment

While this may not be true under all contracts, it is generally understood that the purpose of a price review clause in a long-term gas supply contract is to provide a procedure by which either party may seek to revise the contract price in order to maintain or restore the parties' bargain. Under such price review clauses, this means that the purpose of the adjustment phase is to determine a revision of the contract price appropriate to restore the parties' bargain when the contract price no longer reflects that bargain because of changes in circumstances or market developments (as defined in the particular price review clause).

Price review clauses differ in how they address any standard to be applied in determining whether an adjustment is appropriate, and the relationship between trigger and adjustment.

Some price review clauses are very specific in identifying the relationship between trigger and adjustment. For example, the AIPN Model Form Gas Supply Agreement provides (as one alternative) that, if the seller's costs or taxes increase or its revenues decrease by more than a certain percentage because of a change in the law, the seller is entitled to an adjustment of the contract price to reflect

2 The quoted clause is taken from the judgment in *Gas Natural Aprovisionamientos, SDG, SA v Atlantic LNG Company of Trinidad and Tobago*, 2008 WL 4344525, at *1 (SDNY) as well as Exhibit A to the von Mehren declaration in support of the motion to confirm arbitration, which included further extracts from the Final Award in the UNCITRAL Arbitration, *Atlantic LNG Company of Trinidad and Tobago v Gas Natural Aprovisionamientos, SDG, SA* (17 January 2008) at para 4.

3 See Final Award in ICC Case No 9812, August 1999, "Extracts from ICC Arbitral Awards: Price Setting and Price Revision in the Energy Sector" (2009) 20(2) ICC *International Court of Arbitration Bulletin* 69–70.

the increase in costs or taxes or decrease in revenues so that the financial position of seller remains unaffected.[4]

However, many long-term gas import contracts use broader language to address both trigger and potential adjustments. For example, many long-term gas supply contracts include price review clauses that refer explicitly or implicitly to a "a fair and equitable" revision of the contract price. This formulation can be seen in the *Atlantic LNG* clause quoted above.

The phrase "fair and equitable" is relatively broad, and disputes can and do arise over what it means within the context of a particular contract. Beyond the plain meaning of these terms, there may be some guidance from the applicable law and/or the parties' negotiations and past practices. Some parties may also argue that there is an industry understanding of these terms.

Some clauses do not expressly use the phrase "fair and equitable," but state that the adjustment must "reflect" the change in circumstances that triggered the price revision. This type of language can be seen in the clause in ICC Case No 9812, which states that an adjustment must "reflect such changes". In other words, it expressly links the adjustment to the changes that triggered the need for a revision: in that case, "such changes" referred to the trigger requirement that the "economic circumstances in the country of the Buyer which are beyond the control of the Parties" have "change[d] significantly compared to what is reflected in the prevailing price provisions". Under such a clause, if, for example, a party relied on a change in the end-user value of gas in the market to trigger its revision and that satisfied the contractual trigger, it might have to prove that its proposed adjustment to the contract price provisions "reflects" this change.

Even where the contractual language does not say so expressly, parties often argue that an adjustment must nevertheless be fair and equitable and/or reflect the change in circumstances on which the revision is based. And, regardless of the language of the clause, disputes can arise as to how precisely a party must show that its proposed adjustment addresses the basis for the price revision.

Under any standard, it may be difficult to quantify the effect of changes of circumstances in a market or reflect such a change in a price formula with precision. One of the reasons for this is the nature of the price mechanism in many long-term gas supply contracts. The contract price in many long-term gas supply agreements is calculated on the basis of formulae. Historically (and still today), many long-term gas supply contracts have used contract price formulae composed of a base price indexed to specified commodities (such as prices of alternative fuels, gas hub prices, etc).

Consequently, there may be many ways to adjust the contract price formula to address a change in circumstances, particularly where the alleged change is a

---

4 See AIPN (2006) Model Form of Gas Sales Agreement, at Article 15.9.1 (Alternative 1).

broad change in market circumstances (such as market liberalisation) rather than a development directly affecting one component of the formula (such as one of the competing fuels used in a price index).[5] These might include adjusting the base price, adjusting the indexation components, applying an overall percentage discount or premium to the contract price calculation, or adding an additional component to the contract price formula. As discussed below, in addition to revising the different components of the contract price formula, there also may be an issue as to whether a tribunal can change other terms in the contract, such as the delivery or flexibility provisions, and whether that affects the adjustments the tribunal may make to the contract price mechanism.

The nature of contract price formulae and the various ways they can be adjusted can make it difficult for a requesting party to prove that the specific adjustment proposed is appropriate to reflect the change in circumstances (and for arbitrators to understand and assess the effects of a proposed adjustment). Disputes therefore can arise as to what the requesting party's burden is in order to prove whether a proposed adjustment is an appropriate one.

4. The scope of an adjustment

The scope of a tribunal's authority to adjust contractual terms beyond the contract price is important because the contract price in a long-term gas supply contract usually is intended to reflect all of its terms, including provisions addressing quality, flexibility, take-or-pay, delivery, discounts and penalties. In commercial negotiations between parties, it is often the case that a price adjustment will be agreed in conjunction with adjustments to these other terms. However, an arbitral tribunal may not have the authority to do that.

In *Atlantic LNG*, for example, the clause was silent as to whether the tribunal's authority was limited only to adjusting the contract price provisions. The tribunal considered the scope of its authority and concluded that it did not have the authority to change other contract terms and that this also created potential limits on its authority to adjust the contract price. It explained that it:

> *interprets Article 8.5(f) as authorizing it to revise the price provisions of Article 8, but does not believe that it is empowered to revise any of the other provisions of the Contract.* ***Implicit in this limited approach to this dispute is the obligation to ensure that any revision to the pricing mechanism would not have significant adverse implications under any of the other provisions of the Contract.***[6] Authors' emphasis added.

Issues also can arise in some cases as to whether a tribunal's authority is

5 For example, if the excise tax on gasoil (HEL) is raised, this could affect the calculation of the contract price where gasoil (HEL) is used in an oil-indexed price formula.

6 *Atlantic LNG* Final Award, at para 21.

limited to 'adjusting' the existing contract price formula or whether it has the authority to replace the contract price with an entirely new mechanism or formula, and whether the tribunal can order an adjustment that is different from that requested by either or both parties.

5. Application of any relevant benchmarks

In determining the appropriate adjustment, the parties and the tribunal will need to establish whether there are relevant benchmarks against which to compare the prevailing contract price and, if so, how to apply them. The price review clause may explicitly specify that certain benchmarks are relevant.[7] For example, the clause may refer to import prices at the border of the market at issue or in other markets, or to a level of the market further downstream (eg, wholesale, end-user, etc), or to a number of benchmarks.[8] The clause may also provide that the adjustment should reflect changes in the relevant market, including changes in the value of natural gas.

Some price review clauses give specific guidance with regard to what benchmark data must or should be considered in determining the adjustment. For example, section (b) of the *Atlantic LNG* clause provides that "the Parties shall take into account levels and trends in price of supplies of LNG and Natural Gas [in redacted market] such supplies being sold under commercial contracts currently in force on arm's length terms, and having due regard to all characteristics of such supplies".[9] As discussed below, there are issues of proof that may arise in establishing benchmark prices for use in determining the adjustment, including with respect to the 'characteristics' of supply.

In determining which benchmarks may be applicable, how to calculate them, and how to compare them to the contract price, expert evidence is usually critical. The analyses underlying these determinations involve industry-specific data, and can involve extensive and detailed calculations. The use of expert evidence in this regard is a common feature of many price review arbitrations and can lead to the tribunal needing to make a number of judgments regarding the assumptions and calculations underlying the expert analyses and regarding the reliability of the experts.

5.1 Import prices

Some clauses refer to the price level "at the point of import" as a relevant benchmark. If import prices are to be considered, a tribunal will need to determine what import prices are relevant. Some clauses will refer to specified

---

7 Clauses referring to price levels at designated points are sometimes referred to as 'landscape' clauses.

8 For example, a clause may also refer to some or all of import prices, hub prices, and end-user prices as benchmarks.

9 Section (b) of the *Atlantic LNG* clause is written in mandatory rather than permissive terms. It says that the parties shall take into account the levels and trends in the price of LNG and natural gas, rather than that the parties may take such factors into account.

markets and indicate that only those markets are relevant while others may refer to certain markets without excluding others.

Some clauses specify import prices in one geographic market (eg, 'Spain') while others use broader language (eg, 'Northwest Europe'). This creates scope for dispute as to how to interpret such a phrase. When multiple countries or a region are specified, disputes can arise about whether all the countries should be considered, and how to weight or adjust data from different countries (how data is weighted/adjusted can impact calculations).

When benchmarks are referred to, price review clauses often specify that the characteristics of the gas supplied must be taken into account because different gas products will be priced differently. Such a clause may refer to "having due regard to all characteristics of such supplies (including, but not limited to quality, quantity, interruptability, flexibility of deliveries and term of supply)", as the *Atlantic LNG* clause does. However, there can be disputes about how to take into account characteristics such as quantity, continuity/interruptability, duration, and flexibility when considering publicly available benchmark data that is not segregated by these characteristics.[10]

Price data for specific contracts is typically confidential, but import price data is available for many markets through national statistical authorities, although it is not available for all markets.[11] Even where data is available, there can be a lag in time between the date on which gas is imported into a market and the date on which the statistical authorities publish the average import price. Such data may or may not be revised retroactively to reflect the impact of retroactive price revisions that may have occurred subsequent to the data being reported.

5.2 End-user prices

Some clauses refer to end-user prices as a benchmark. For example, the clause in ICC Case No 9812 does not refer to import prices, but it directs the parties to consider data with regard to end-user value in the buyer's market, referring to "an adjustment of the price provisions ... reflecting such changes, in particular the value of Natural Gas in the end user market of the buyer as such value can be obtained by a prudent and efficient gas company".

The award in ICC Case No 9812 describes the use in that case of an analysis that attempted to approximate "the value of Natural Gas in the end user market of the buyer as such value can be obtained by a prudent and efficient gas

10 The quality of the specific gas supplied under a contract is not usually an issue in dispute at the adjustment phase. Most gas delivered through pipelines (or as LNG cargoes that are regasified and then delivered through pipelines) is high-calorific gas and once gas is in the pipeline and taken by a buyer, the individual gas molecules are interchangeable.

11 In Europe, for example, various national authorities publish such data, as does Eurostat, the statistical agency for the European Union. For the US, such data is available through the US Energy Information Administration.

company". The approach taken there looked to the gas market value, described below. Another potential approach might be to look at the prices paid by the final consumer of gas "at the burner tip" and then subtract the costs that were incurred in the distribution chain between the final consumer and the gas importer.[12]

There are logistical complexities that may complicate any analysis of end-user (or other) prices in a market. These include the availability of data, the lag in time between the date on which end users pay for the gas and the time in which statistical authorities publish data reporting such prices, and the specificity or granularity of the available data.[13] Disputes may also arise regarding how to calculate the various costs in the gas distribution chain between the import level and the point of consumption (including costs of storage, structuring, etc) and how to attribute those costs to a "prudent and efficient gas company".

5.3 Hub prices

Some clauses refer to prices from gas hubs as a benchmark. For example, a clause may say that the parties should take into account "the price level at gas hubs in Northwest Europe that are generally considered to be liquid (including NBP, TTF, Zeebrugge, etc)". Other clauses may not contain a direct reference to hubs, but rather may refer, as in the *Atlantic LNG* clause, to "levels and trends in price of supplies of LNG and Natural Gas" in a particular market or markets, in which case parties may consider using hubs as a benchmark.

Hub prices are readily available and are published daily. However, many markets do not have gas hubs or even access to gas hubs due to pipeline infrastructure constraints,[14] and disputes may arise as to whether a hub is sufficiently liquid to serve as a relevant benchmark. Consequently, whether hub prices are a relevant benchmark for an adjustment (and what hub prices are relevant) will depend on the relevant market and what the price review clause provides.

---

12 For example, although the underlying contract in ICC Case No 9812 is the same as that in ICC Case No 13504, the two cases reflect different approaches to ascertaining end-user value. The tribunal in ICC Case No 9812 employed a gas market value approach, while the tribunal in ICC Case No 13504 referred to the end-user value of gas as encompassing prices actually obtained in the various end-user market segments. See Final Award in ICC Case No 13504, (2009) 20(2) *ICC International Court of Arbitration Bulletin* 100–02.

13 Disputes may arise, for example, in connection with whether the end-user data at issue is divided into all relevant end-user segments such as residential, commercial, industrial, and power, and whether a weighted average will be used; whether data is published monthly or only quarterly or annually; and, if annual data is published, how monthly or quarterly data may be derived from it.

14 For example, there are not yet gas hubs in Asia. Markets in Asia have not yet liberalised, and price formation at the wholesale level is predominantly based on oil products.

5.4 Use of forward-looking or backward-looking data

One issue that sometimes arises at the adjustment phase is what time period should be considered when determining whether a proposed adjustment is appropriate. In other words, for purposes of an adjustment, should data be considered from the period in which the adjusted price will apply even though that data was not available at the time when the price review was requested?

Parties may argue that determining whether the trigger requirements for a change in circumstance or developments in the market have been met is a backward-looking exercise: it looks backwards in time in order to determine whether the change in circumstances or other market developments stipulated in the price review clause occurred in the relevant time period.

By contrast, parties may argue that the adjustment stage of a price review requires a forward-looking analysis. Once a price revision is agreed or ordered by a tribunal, the revised price usually is effective (retroactively) from the date on which the requesting party formally notified its price review request (or some other date stipulated by the parties). This means that, during the course of the arbitration, there is often data available about both the development of the contract price and the development of benchmark prices that post-dates the price review request. In some cases, in the course of the arbitration, there may be several years of data about the period for which the proposed adjustment will retroactively apply.

This can be illustrated by the timeline of a hypothetical arbitration in which a party issues a price revision request notice on 1 January 2016. In this hypothetical scenario, the parties have agreed that any revision will be retroactive to the 1 January 2016 date of that request. The hypothetical contract requires the parties to negotiate for six months before arbitration can be commenced. The parties negotiate until 1 July 2016 but do not reach agreement. The parties then extend their negotiations for another three months, but again do not reach agreement, and arbitration is commenced in October 2016. The evidentiary final hearing takes place in December 2017, 15 months after the commencement of arbitration, and the award will be rendered more than two years after the effective date of any adjustment is ordered by the tribunal.

In this scenario, during the arbitration, the parties and the tribunal may be able to see what the effect of the proposed adjustment would be during at least part of the period in which it will apply (ie, for at least part of the period after 1 January 2016). This may raise issues as to what evidence the tribunal is to consider in determining whether a proposed adjustment is appropriate.

6. "In any case" clauses

Some contracts have what is often referred to as an "in any case" clause. There are various formulations of these types of clauses. *Atlantic LNG* and ICC Case No

9812 both have forms of "in any case" clauses. Some, like the clause in ICC Case No 9812, provide that a buyer must be able to "economically market" the gas sold under the contract while others, like the *Atlantic LNG* clause, provide that the contract price "[s]hall allow the Buyer to achieve a reasonable rate of return" on the gas supplied under the contract. Others may refer to the ability to sell the gas delivered under the contract in competition with other sources of energy in the market (sometimes stipulating a particular market segment).

There are various disputes that can arise when a buyer relies on an "in any case" clause as a defence to an upward price adjustment for the seller. These include determining the buyer's market. This may raise issues as to what market segment the buyer competes in. It also may be complicated if the buyer re-exports gas to other markets.

When a clause refers to other sources of energy or 'competing' forms of energy, it can also lead to disputes about identifying those sources and whether and which other sources can be said to 'compete' with the gas sold under the relevant contract. Some such clauses specify the fuels to be considered (and some specify that natural gas is to be considered while some exclude natural gas as a competing fuel), and some are silent in this regard. This can have a significant impact on the analysis. Where gas sold under the contract is compared to other forms of energy (such as coal, fuel oil, wood pellets, etc) at the point of consumption, one approach may be to apply a detailed analysis known as a "gas market value", which makes a comparison of natural gas to other sources of energy in the market.[15]

Determining if and how to apply an "in any case" clause may also raise issues about whether evidence must be introduced about the buyer's business practices, including its actual sales, in order to assess a claim that it cannot economically market the gas if the price is increased or what a "reasonable rate of return" is.

7. Procedural issues

In addition to some of the evidentiary issues discussed above, there are a number of other procedural issues that may arise in connection with the adjustment phase of a price review. These issues range from when an adjustment must be proposed to when adjustment issues are to be considered during an arbitration (eg, whether, in the particular circumstances of a case, trigger and adjustment should be considered together or whether the issues are bifurcated so that adjustment is only considered if the trigger requirements have been found).

---

15 Although it undertook such an analysis in another context, the tribunal in ICC Case No 9812 described it as follows: "[w]hen calculating the market value of natural gas, the capital and operating costs for gas and the alternative fuel, as well as the fuel costs of the alternative fuel, have to be considered. The theoretically affordable gas cost is the mathematical difference between on the one hand the sum of the capital, operating and fuel costs for the alternative fuel and, on the other hand, the sum of the operating and capital costs for gas". Final Award in ICC Case No 9812, at p75.

One issue that can arise is whether a party can change the adjustment it proposes during the price review process, including during an arbitration. As with other issues, this will depend on what the parties agreed in their price review clause, but it is not unusual for there to be disputes about changes that a party has made to its proposed adjustment during the price review process.

These issues can be complicated by the fact that there may be competing requests for revision; that is, the buyer may be seeking a downward price adjustment while the seller is seeking an upward adjustment, based on the same or different alleged contractual triggers. Moreover, even where the parties do not have competing requests, a party opposing a price revision will usually need to make the strategic decision whether it should nevertheless propose an alternative adjustment in the event that a tribunal determines the trigger has been met.

The result of the adjustment phase often depends on the ability of an arbitral tribunal comprised of lawyers to assess and apply expert evidence. There may therefore also be issues about what expert assistance should be available to the tribunal in making a final decision on an appropriate adjustment, particularly where the tribunal has to consider different and/or multiple proposed adjustments.

8. Conclusion

The adjustment phase of a gas price review arbitration usually involves complex and industry specific issues of pricing and economic analysis, including with respect to determining what benchmarks, if any, are relevant. As with other aspects of gas price reviews, the approach taken will vary depending on the specifics of the case, the language of the contract, and the applicable law. These factors must be carefully considered and assessed in determining an appropriate adjustment to the contract price, and to ensure that an adjustment is consistent with the parties' bargain.

Natural gas price review arbitrations: issues in adopting hub indexation

Luis Agosti
Boaz Moselle
Compass Lexecon

1. Introduction

In this chapter we discuss some issues that are typically 'live' in price reviews arising from longterm natural gas supply contracts where the markets supplied are liberalised and downstream pricing is determined by the prices set through the trading of natural gas as a commodity ('hub trading'). In practice this means a predominant (though not exclusive) focus on Northwest Europe. Other parts of Europe may follow, and in the future potentially Asian markets, though that remains a more distant prospect. In our experience moreover, the issues we discuss have to date usually been more germane to contracts for the delivery of piped gas than of LNG (in part because most of the markets in question receive relatively little LNG, but also because LNG pricing can also be influenced by the global LNG market). Below we will for the most part refer to gas, which can be understood as a reference to piped gas, though much of what we say will also be potentially applicable to LNG.

As a preliminary caveat, we note that this is a general discussion, not an expert report. In the interests of space and readability, we therefore eschew excessive detail. Any application to an individual contract or price review will of course depend on the specific legal, factual and economic matrix, which may create circumstances that would modify the general conclusions presented here.

In the first decade of this century (and before) price reviews were generally related to the adjustment of oil-indexed price formulas in a market context of little or no competition.[1] Oil-indexed formulas were adjusted, typically, by simply raising or lowering the price level while retaining the structure of the oil-indexed price formula. To give a simplified example, one might have had a contract with a formula along the following lines:

$$P_{gas,t} = P_0 + B \times (P_{oil,t} - P_{oil,0}) \quad (1)$$

1 In this chapter we use 'oil' as shorthand for what is sometimes crude oil but often a basket of fuels such as refined oil products (gasoil, fuel oil) and possibly also coal.

where $P_{gas,t}$ is the price at time *t* of the natural gas supplied under the contract, $P_{oil,t}$ is some measure of the price of oil at time *t*, and $P_{oil,0}$ is the value of that measure at some starting point labelled time 0.[2] A price review of the kind we are describing might then have changed this constant term, but left the 'slope' in the formula unaltered.

However, many European gas markets have changed significantly since 2000, with the development of 'gas-to-gas competition', as we explain briefly below. In particular, this meant that market participants at the wholesale level in many countries increasingly bought and sold gas at 'trading hubs', creating reliable, publicly available market prices of gas traded over various time frames ('hub prices').[3]

These developments have in turn led to a more profound adjustment of pricing arrangements in many long-term supply agreements, with the price development no longer driven by oil prices but by indexation to hub prices. Continuing the example above, the formula shown in (1) might be replaced by the following:

$$P_{gas,t} = P0 + (P_{hub,t} - P_{hub,0}) \qquad (2)$$

where $P_{hub,t}$ is some measure of the price of gas sold at a European hub, and $P_{hub,0}$ is the value of that measure at time 0. Since P_{gas}, $P_{hub,\ t}$, and $P_{hub,0}$ all refer to the price of gas, there are generally no pass-through factors (ie, the 'B' factor in the formula shown in (1)) in a formula with indexation to hub prices.

Gas markets in many European countries now provide liquid and transparent price signals through gas hubs that reflect supply/demand fundamentals. As a result, a large share of long-term gas supply agreements in Europe is now partially or fully indexed to European hub prices. The International Gas Union (IGU) reported that in 2017 gas-to-gas competition determined the price for 70% of wholesale volumes on average in Europe, 92% if the focus is on Northwest Europe.[4] Markets in Central Europe, that in some

---

2 To further elaborate, P_0 is the price at the starting point time 0, sometimes called the 'base price'; the expression in brackets $(P_{oil,t} - P_{oil,0})$ is the change in oil price between the starting point and time *t* (which may be positive or negative); and the formula sets the price at time *t* equal to the price at time 0 plus some multiple B applied to the change in oil prices since the starting point. B is often called the 'slope' term of the formula and indicates to what extent changes in the price of oil are reflected in the price of gas. If the gas and oil prices are in the same units then B is referred to as the 'pass-through factor' (and will often be less than 1).

3 Several sources provide detailed descriptions of this transition, including: P Heather, *The evolution of European traded gas hubs* (The Oxford Institute for Energy Studies ('OIES'), December 2015); J Stern and H Rogers, *The Dynamics of a Liberalised European Gas Market – Key determinants of hub prices, and roles and risks of major players* (OIES, December 2014). Note that as with any document cited in this article, citation does not imply unqualified endorsement of content.

4 See IGU Wholesale Survey (IGU, 2018) pp13–14. Northwest Europe refers to Belgium, Denmark, France, Germany, Ireland, Luxembourg, Netherlands and UK. For imports, the IGU defines wholesale prices as the price at the "first point of sale", eg, if gas is first sold at the border of a country under an oil-indexed contract and it is then re-traded at the hub (and therefore at hub prices), the IGU considers these volumes as oil-indexed volumes.

5 IGU Wholesale Gas Price Survey (IGU, 2018) p14. IGU's definition of Central Europe comprises Austria, Czech Republic, Hungary, Poland, Slovakia, and Switzerland.

cases have less intense gas-to-gas competition, also exhibit a high degree of hub indexation, at 73%.[5] Only in relatively few countries (eg, Spain, Portugal and Turkey) does oil indexation prevail as the dominant pricing mechanism.

1.1 Implications for price reviews

Most long-term gas contracts were agreed at a time when there was no wholesale market for gas, and consequently no visible, reliable wholesale gas price to which the contract price could be indexed.[6] Gas in long-term contracts was often priced against the price of alternative competing fuels such as oil or coal (or those prices were used to index the gas price, even without explicit considerations of inter-fuel competition). However, that pricing mechanism had certain imperfections. For example, the development of competing fuel prices did not always reflect changes in the gas market, in average border prices, in the 'value of natural gas' or in other applicable criteria in the way that the parties might have originally wished or expected. Price reviews were thus introduced in long-term contracts as a mechanism to reset contract prices pursuant to parties' agreed criteria.[7]

In theory, the development of liquid hubs and the subsequent indexation of prices in long-term supply agreements means a much more limited role for price reviews. Since hub indexation eliminates (or at least greatly reduces) the risk of mismatches between prices in long-term import contracts and market prices of natural gas (which are now reflected in hub prices), the need for price review provisions would be reduced. Indeed, it is well known that at least one major producer, Equinor in Norway,[8] has agreed with many of its customers to replace long-term oil-indexed contracts with new contracts that are hub-indexed and that no longer contain the type of price review provisions described above.

The reality, however, is far more complex. We continue to see price review disputes in arbitration, even in gas supply contracts that are already indexed to hub prices or where the adoption of hub-indexation is no longer part of the discussion. The issues are of course different from the issues that arise in contracts with oil-indexed formulas and mainly have to do with two factors. The first is price level. Buyers and sellers have different views on the appropriate discount or premium relative to hub prices that long-term contract prices

---

6 Import prices derived from customs declarations, and in particular import prices reported by the German Federal Office for Economic Affairs and Export Control (BAFA), have often been used in negotiations as a benchmark for prices in long-term import contracts. However, these prices have not been widely used as a price index in the price formulas of long-term import contracts, among other reasons because they are only available some months after delivery.

7 Anthony J Melling, *Natural gas pricing and its future: Europe as the battleground* (Carnegie Endowment for International Peace, 2010) p19. Of course, each price review clause is a specific contractual provision, individually negotiated between the parties and to be understood in the context of the contract as a whole and taking into account a range of potentially relevant matters (applicable law, history of contract negotiation, history of performance, etc). In our experience there is no 'industry standard' type of provision in these matters.

8 L Franza, *Long-term gas import contracts in Europe* (Clingendael International Energy Program, 2014).

should reflect.[9] Secondly, buyers and sellers have different views on the right indexation structure, including, the type of contracts ('hub products') that drive the changes in the contract price level over time.

In this chapter we address these 'open' issues. In section 2 we describe what hub prices are and how hub indexation works. In section 3 we look at the interaction between resetting the price level as of an effective date and the adoption of hub indexation. In section 4 we discuss the appropriate hub indexation structure. We also address other issues that are relevant from an implementation perspective but are generally less likely to be contentious. Elsewhere in this book, Valentina Bonetti, Dan Harris and Carlos Lapuerta also focus on certain potential sources of dispute between parties to long-term contracts after they have adopted hub price indexation, namely transportation costs, market risk, margins and the risk of market manipulation. In this chapter we too discuss issues related to margins and market risk, although with a somewhat different focus from those authors.

As a final remark, we note that this article is about hub indexation in the context of long-term contracts and price reviews. We do not discuss hub indexation in new supply contracts, or in renegotiations that take place outside of the context of price reviews, where changes in pricing arrangements are not bound by the same limitations and conditions that parties to long-term contracts (and arbitral tribunals resolving price review disputes) face in price reviews.

2. Hub contracts, hub indexation and differences with long-term contracts

A hub can be thought of as a central pricing point for natural gas, where gas is traded via a set of standardised contracts and the price of the gas is set through the interaction of supply (sellers) and demand (buyers). A hub can be either a virtual location (eg, a market area that corresponds to the borders of a region or country, such as the National Balancing Point ('NBP') hub in the Great Britain) or a physical point on a gas network (eg, an interconnection point where gas pipelines meet, such as the Henry Hub in Louisiana, US). Other hubs in Europe include the Title Transfer Facility ('TTF') in the Netherlands, the Zeebrugge ('ZEE') and Zeebrugge Trading Point ('ZTP') hubs in Belgium, the NetConnect Germany ('NCG') and GASPOOL ('GPL') hubs in Germany, and the Gas Exchange Point ('PEG') in France.

Trading at gas hubs can be undertaken through:

- over-the-counter ('OTC') trading, in which two parties trade bilaterally, usually using a broker or dealer to facilitate trades, under standard terms; or

9 When the price in a long-term contract is indexed to a hub price, the contract price moves in parallel with the hub price. However, it may be set above (premium) or below (discount) the hub price. Hub indexation therefore does not mean, necessarily, hub price levels.

- trading on a commodity exchange, that allows for anonymised trading under standardised terms.

A standard hub contract is a contract for the delivery of a fixed volume of natural gas, at a constant rate, over a defined period in the future, at a virtual or physical geographical location within a gas system. For example, a standard TTF month-ahead contract traded on 15 March is a contract for the constant delivery of 1 MWh per hour of natural gas over the following month of April at the TTF hub in the Netherlands.[10] Different contracts may allow for settlement to take place 'physically' (ie, by the delivery of the specified gas in line with the applicable contractual provisions), and/or 'financially' (ie, by one party making a payment to the other that represents the difference between the price agreed in the contract and the applicable price of gas at the time of delivery).

Depending on the level of maturity of the hub,[11] different products may be available. Less mature hubs may offer products for spot or prompt delivery within the day, day-ahead or month-ahead.[12] More developed and liquid hubs would also offer forward products, that is, contracts for delivery in the more distant future, for example, quarter-ahead, season-ahead (summer/winter) and multi-year-ahead of the trading date. In this sense, there are substantial differences between the level of liquidity and traded volumes in different hubs and for different hub products. The TTF is now the largest hub in terms of traded volumes and is also the most liquid, ahead of the NBP in the UK. In terms of products, trading contracts for spot and prompt delivery is generally more liquid than trading for forward products.[13] We discuss the relevance of the different hub products in section 4.

In the context of long-term gas supply agreements, hub indexation simply means that the contract price follows changes in the price of hub product(s) (rather than changes in the price of oil or other commodities or indices). There are however important differences in terms of how oil-indexed prices and hub-indexed prices allocate price risk.

To explain these differences, we first refer the reader back to the two price formulas shown earlier. In purely mechanical terms, the two formulas work in

---

10 The standard agreement by the European Federation of Energy Traders ('EFET') is for 'Flat Transactions' where the contract quantity is the same for each hour (unless otherwise agreed by the parties). See *General Agreement Concerning The Delivery And Acceptance of Natural Gas*, EFET, May 2007, and the corresponding annexes for each hub.

11 A number of publications discuss the state of development of different European hubs, including for example *European Gas Hub Development Study*, EFET, 12 August 2014, and *Results of Monitoring the Internal Electricity and Natural Gas Markets in 2017, Gas Wholesale Markets Volume*, ACER, October 2018.

12 'Spot' and 'prompt' generally refer to contracts for delivery in the short-term, eg, within the day, day-ahead, week-ahead, etc. up to month-ahead delivery. We note however that there is not a unique classification of prompt and forward contracts. Month-ahead contracts may be considered as forward contracts by some authorities.

13 *Annual Report on the Results of Monitoring the Internal Electricity and Natural Gas Markets in 2017, Gas Wholesale Markets Volume*, ACER, October 2018.

a similar manner. They both establish an initial price level, (that may, for example, correspond to the price at the date on which the buyer and the seller in a price review negotiation agree that the revised contract price will first apply). The price in the following periods is then equal to P_0 plus a change determined by the change in the price of the commodity used for indexation purposes: the price of oil in the case of the oil-indexed formulas and the hub product price in the case of hub-indexed formulas. A difference is that in the case of the oil-indexed formula, only a fraction of the change in the price of oil translates to the change in the contract price because of the pass-through factor *B*; while for hub-indexed contracts it is usual to pass-through all of the change.

This subtle difference has important implications for the sharing of price risk between the buyer and the seller. Price risk can be described as the impact on a party's cash flows and profits arising from variations in the contract price of gas supplied under the contract. Under a long-term contract, a seller faces price risk, as there is little or no link between its upstream costs of production and the price it achieves from selling the gas (whether that is indexed to oil products or to hub prices). A buyer faces price risk to the extent that the contract price it pays for gas differs from the price it can achieve in its downstream sales.

In the past, when the price of oil went up, the buyer would be able to resell its gas for a higher price (in the short run, because it resold gas at oil-indexed prices; and in the long run, because gas was often sold in competition with oil products). However, it would also have to pay a higher price. The net effect on its profitability depended on the difference between: (a) the increase in sales price it could achieve; and (b) the increase in cost due to a higher purchase price under its long-term supply contracts.

The introduction of pass-through factors with values most often in the 80% to 90% range[14] meant that the latter factor was smaller than it would have been with a 100% or higher pass-through factor. The buyer would therefore obtain a higher margin than if 100% of change in the price of alternative fuels was translated to the contract price. Similarly, if the price of alternative fuels decreased, the contract price would decrease by a lesser amount, decreasing the margin the buyer obtained when selling the gas downstream.

Pass-through factors less than one therefore passed a greater part of the price risk to the buyer, by giving it bigger swings in profitability in response to changes in oil prices than would have been the case with no pass-through factors (equivalently, with pass-through factors equal to 1).

On the contrary, in competitive and liquid gas markets the hub price is already a wholesale market price, which generally also determines gas prices in

14 See *Putting a price on energy: international pricing mechanisms for oil and gas*, Energy Charter Secretariat (2007) 154.

the end user market. If the hub price increases by €1/MWh then in general the resale price will increase by €1/MWh, and the price in the supply contract, if it is hub-indexed along the lines of formula (2) above, increases by €1/MWh.[15] The buyer's margin on resale is therefore unaffected: the buyer is no longer subject to the risk of the contract price decoupling from the prices it can obtain when reselling the gas it buys under the long-term contract.

The fact that the buyer is no longer subject to price risk does not mean that it is always going to make a margin. If P_0 is lower than the hub price at that time P_{hub0}, the buyer will obtain a positive margin equal to at least the difference between the initial contract price and the hub price and higher if it can sell downstream to end users at prices above hub. If P_0 is higher than P_{hub0}, then margins will be lower, though not necessarily negative. We explain below that long-term contracts provide buyers with valuable flexibility. This flexibility may allow buyers to obtain resale prices above hub levels and can thus justify having contract prices at a premium to hub prices.

Therefore, a key question in the determination of a hub-indexed contract price is the existence and magnitude of a premium/discount relative to the hub price itself.

3. Resetting the price level in a hub-indexed contract

In the context of a price review process, resetting the contract price level of course requires consideration of the contractually specified criteria. For our purposes it is sufficient to note that different contracts contain different criteria, which can lead to quite different outcomes. In particular, there is no reason to think that the criteria will lead to the adoption of a price as of the effective date that is equal to 'the hub price'. Most fundamentally, the criteria of most price review clauses are not directly related to hub pricing, and are unlikely to lead to a price level exactly equal to a hub price. Quite apart from this contractual logic, it is also important to note that there is no such thing as a single hub price: there are many prices depending on the specific hub chosen (eg, TTF, PEG, etc) and the specific 'hub product' (eg, contracts for delivery day-ahead, month-ahead, quarter-ahead, etc).

For similar reasons, there is no general rule whereby a price review should lead to the adoption of contract prices at a premium or at a discount relative to hub prices. Buyers will generally argue that a discount is required to market the gas economically (which for simplicity can here be taken to mean profitably, or without incurring a loss). Sellers will argue, on the contrary, that a discount to hub represents "free money", since the buyer can simply resell at the hub with

15 Here we abstract away from issues of timing (e.g, if the buyer has agreed resale at a hub price fixed at a certain point in time), and also of hedging (which the buyer would typically use in those circumstances). Our point remains valid in principle, but the details become more complex, when one takes those issues into account.

a guaranteed profit; and therefore that it would significantly alter the risk allocation in the contract, giving the buyer would be able to make a guaranteed margin without incurring any risks.

In our experience, different price review criteria will lead to different outcomes. For example, while some price review clauses do state that buyers should be able to market the gas economically (subject to certain conditions), other type of price review clauses do not include such criteria. In addition, even in those cases in which the price review clause says the buyer has to be able to economically market the gas, that does not per se justify a discount to hub prices because retail prices are generally set at a premium to hub prices. In this case, whether a discount or a premium is required will depend on the conditions of each market, as well as other factors (eg, other contractual price review criteria).

One particular issue is that (absent clear contractual provisions to the contrary) the revised price as of the effective date will need to recognise the value of the non-price terms of the contract in question. While there are many non-price differences between long-term contracts and hub contracts (eg, term, *force majeure* credit requirements, etc), the most salient issue in our experience is the flexibility in the contract. As we now proceed to explain, long-term contracts usually contain flexible supply terms that create value for the buyer relative to inflexible hub contracts, and therefore merit a premium relative to the hub price, all else being equal.

Long-term contracts are flexible because, generally, they allow the buyer to decide, within certain limitations, when and how much gas they off-take within a given period. By contrast, standard hub contracts specify 'flat' delivery (meaning that the volumes traded are delivered at a constant 'flow rate' during the whole of the delivery period).

Buyers need this flexibility (from somewhere, not necessarily from their supply contracts) because the demand they serve is not constant over time. Gas consumption is higher in the winter and lower in the summer; and, at a more granular level, is higher on (for example) cold days than on hot days (in countries where gas is used for space heating). The flexibility is valuable because in its absence buyers would need to incur storage costs or buy and sell gas at the hub to accommodate their off-take profile to the demand they serve.

Two examples illustrate the value of flexibility. First, consider a market in which 70% of the demand for Year 1 takes place during the winter and 30% takes place during the summer. There are two companies A and B that serve this market. Company A has a year-ahead hub product for delivery of gas in Year 1 at a price of €10/MWh. This year-ahead contract is a flat contract that delivers the same amount of gas in the winter and in the summer. The price of €10/MWh is the average of the price of the hub contract in the winter, €12/MWh, and the price of the hub contract during the summer, €8/MWh. Company B has a long-term contract that the parties agree to set equal to the price of the year-ahead contract for Year 1, also €10/MWh. The long-term contract, however, is flexible and allows the buyer to take any amount of gas during the winter and the summer season subject to a minimum total take over the year.[16] The flexibility in the long-term contract is then valuable for Company B, which can offtake 70% of its contract quantity in the winter and 30% in the summer without affecting its costs. Company A on the contrary has to off-take the same amount of gas in the winter and in the summer. To cover its customers' demand, however, Company A will need to buy additional gas from the hub during the winter: it receives 50 units but needs 70, so will need an additional 20 units, at the winter price of €12/MWh. Conversely, it will need to sell the extra 20 units of gas it does not need during the summer (when it receives 50 but needs 30), at the summer price of €8/MWh. The overall effect is equivalent to an additional cost, for every MWh of gas bought under the contract, of €0.8/MWh.[17]

In the example above, the value of flexibility is €0.8/MWh. This is the amount that the buyer under the long-term contract saves in comparison with a year-ahead hub product that has the same average price. The appropriate price adjustment to make the long-term contract price equivalent to the hub price is therefore, all else being equal, a premium on top of the hub price of €0.8/MWh to recognise the value of flexibility the buyer receives and would otherwise have to pay for in the market. In a price review that sought to benchmark Company B's contract against Company A's, one would have to take the price of Company A's contract, €10/MWh, and then add on that €0.8/MWh premium.

In that example, Company B's long-term contract had a fixed price; while that is probably unrealistic, it serves to illustrate the point. However, suppose now an alternative scenario, where the buyer and the seller agree to index the long-term contract price to a hub price. They could agree to a contract price that in the winter is equal to the average hub price during the winter, and in the summer is equal to the average hub price during the summer. This means that when Company B chooses to take more gas in winter and less in summer it automatically incurs a cost: taking 70 MWh in winter and 30 MWh in summer results in an average price of €10.8/MWh.[18] Of course, that would be equivalent to having a long-term contract equal to the price of the year-ahead hub product plus €0.8/MWh.[19]

---

16 For simplicity, assume that the total annual volume is fixed by contract (ie, this is a 100% take-or-pay contract) and that the winter and summer can be taken to each occupy six months. We note that some contracts also specify a daily minimum offtake. This requirement may imply a minimum percentage to be taken in any season (or any six-month period). The intuition is the same even if the minimum percentage of the annual quantity that can be taken in the summer due to daily minimum offtake is, for example, 40%.

17 If the contract is for 100MWh then the cost of buying 20% of the gas at €12/MWh and selling the same amount during the summer at a price of €8/MWh = (20 × 12) – (20 × 8) = €80, which spread out over the 100MWh of gas supplied is €0.8/MWh.

18 The average is (70 × €12/MWh + 30 × €8/MWh)/100 = €10.8/MWh.

19 As discussed below, a contract with indexation to the hub product on a semi-annual basis may still provide valuable flexibility compared to hub contracts.

Consider now a second example, still with the same hypothetical companies and contracts. Suppose that Company B is a utility that supplies residential consumers who use gas to heat their homes. Suppose also that Company B today looks at tomorrow's weather forecast and sees that the day is expected to be unusually cold. Under a typical long-term contract, it can nominate a higher than normal amount of gas to be delivered tomorrow, in anticipation of unusually high demand due to the low temperatures. Unless the price in that contract is indexed to the day-ahead hub price, Company B will not pay a higher price for the additional volumes of gas it purchases to cope with tomorrow's weather-induced peak in demand. By contrast, Company A does not have such contractual flexibility. If it has gas in storage, then it can withdraw some of that gas. However, renting and using storage give rise to additional costs. Alternatively, it could purchase additional volumes of gas at the hub – but the hub price will be elevated, reflecting the unusually high demand for gas tomorrow.

In these two examples, therefore, we see that indexation to the hub product on a semi-annual basis compensates the seller for providing seasonal flexibility (the right to take more gas in winter than in summer), but not for providing flexibility within a season. If Company B's contract price is to be set via benchmarking against hub, and is hub-indexed on a semi-annual basis, then no premium is required to reflect the value of the seasonal flexibility (as explained in the first example), but a premium is required to reflect the value of the within-season flexibility (as per the second example).

The flexibility premium that long-term contracts demand on top of hub prices therefore varies depending on the type of hub product that is used for indexation purposes. The premium will be larger if the contract price is indexed to contracts with a longer duration (eg, year-ahead). In these contracts, the premium will be higher since the buyer will be able to allocate its off-take in periods when demand is higher, and therefore the price is also higher, without affecting its costs. On the contrary, if the long-term contract is indexed to hub products of a shorter duration (eg, season-ahead, month-ahead or even day-ahead), the buyer will still be able to off-take gas during the higher demand period but that will be reflected in the higher prices that the buyer pays the seller in the long-term contract.

In the context of commercial negotiations various approaches have been adopted to dealing with the value of flexibility. In some instances, buyers and sellers have agreed to remove all the contractual flexibility and set the contract price at the level of the hub price with no premium. In the example above of Company B buying gas under a long-term contract, this approach would involve reducing the contractual quantity and selling gas on a flat basis (ie, with the same volume taken each day (or each hour)). The seller would keep the flexibility for itself, in the sense that it would be free to sell the gas that it was no longer supplying to Company B, and could (and rationally would) choose to sell it at times when prices were most favourable.

From the example above, it may seem that sellers and buyers should be indifferent between different pricing alternatives if they all result in the same average price for the seller (or in the same average cost for the buyer). The reality, however, is more complex because flexibility is difficult to value. The type of flexibility that we describe in the example above is only part of the value of flexibility that long-term contracts provide. It reflects the value of flexibility for a set of hub forward prices at a given point in time. This is what economists refer to as the intrinsic value of flexibility. However, hub contract prices change every day and buyers can also derive additional extrinsic value by adapting their off-take profile in response to those changes. For example, as noted above, there is value in being able to procure additional gas, at a given price, in response to 'price spikes' on unusually cold days. How much value depends on the contract price, and the probability of seeing price spikes of various magnitudes. This makes flexibility difficult to value in practice. Moreover, in the context of a price review dispute in arbitration, the probability that a buyer and seller agree on the value of the flexibility in the contract is low.

Despite the difficulty in valuing flexibility there is no disagreement that long-term contracts are flexible products that, all else being equal, command a premium above flat hub products.[20] The value of this premium depends on the type of hub contracts used as a reference in the contract price formula, the premium being higher the longer is the duration of the hub contracts considered.

4. Getting the indexation 'right'

In our experience, parties may also disagree in respect of the indexation that the revised contract formula should reflect. There is not one unique hub in Europe. The UK, France, Belgium, Italy, Germany, the Netherlands and other countries all have their own hubs. Buyers and sellers may have different views as to which hub, or mix of hubs, is appropriate for indexation purposes.

As explained previously, neither is there one unique hub product. Trading encompasses a range of hub products with different durations. Leaving aside the value of flexibility and the trade-off between the premium and the duration of hub products described in the previous section, buyers and sellers may also have different preferences as to the type of hub products they want to consider for other reasons.

20 We note that both buyers and sellers tend to argue that all else is not equal, and that other features of long-term contracts make them less or more attractive than hub contracts.

4.1 Choice of hub

Key issues that may affect the choice of the hub used for indexation purposes include: (i) where the buyer resells the gas, (ii) the liquidity of trading at the hub.

(a) *Where gas is sold*

For many buyers, all else being equal (which is by no means always the case) it makes sense commercially to reference the long-term contract price to the price of the hub in the market in which the buyer is re-selling the gas delivered under the long-term contract. This reduces the risk of contract prices deviating from the prices it obtains downstream.

There are potentially some problems with this approach. First, it may not be possible to identify "the market in which the buyer is reselling the gas". Many buyers buy gas from multiple sources and sell in multiple geographies. Gas is fungible, and there is no unique link between a given customer and a given supply source.

Secondly, there may be no liquid hub in the buyer's market or region. Both parties will typically wish to index to hubs with broad liquidity (*inter alia* for hedging purposes, as explained in more detail below).

Thirdly, while that approach may make sense for the buyer, it may be less desirable from a seller perspective. A seller may wish to index the long-term contract price to one or more hubs that are more representative of average wholesale gas prices in Europe. For example, they may wish to index to the TTF hub, which is the most liquid European hub, rather than take on the specific risks related to the hub in the market of the buyer.

(b) *The liquidity of trading at the hub*

There is no single definition of liquidity. For the purpose of this chapter, it is useful to look at a definition provided by the British energy regulator, the Office of Gas and Electricity Markets (Ofgem): "[t]he ability to quickly buy or sell a commodity without a significant change in its price and without incurring significant transaction costs."[21] There are several measures of liquidity along different dimensions including volume, transaction costs and time:

- Volumetric measures of liquidity include the volume of gas traded, the 'market depth' (ie, the volume of trades in one direction that the market can absorb without materially affecting the price) and the number of active traders.
- A commonly used measure of transaction costs is the bid-ask spread. In natural gas trading markets (and in many other commodity markets) buyers submit 'bids' to buy at a certain price, and sellers submit offers

21 Ofgem, *Liquidity in the GB wholesale energy markets*, 8 June 2009.

('asks') to sell at a certain price. At any point in time, the ask prices are generally higher than the bid prices. To conclude a trade, a seller can accept a bid price, or a buyer can accept an ask price. The difference between the bid and ask price at any point is called the 'bid-ask spread'. A lower spread facilitates transactions and increases the certainty of the price level.

- Time measures refer to how far in advance it is possible to trade gas for delivery at a given point in time. Less liquid hubs tend to offer products for prompt trading. More mature hubs offer products up to four and five years ahead of delivery. Although this is not a direct measure of the liquidity of a hub contracts, it is indicative of the level of maturity of a hub and tends to be correlated to some extent with other measures of liquidity.

Sellers and/or buyers may wish to index prices to a more liquid hub, because they would like to be able to 'hedge' their risks using forward contracts traded at the hub. For example, suppose a buyer has a long-term contract to buy gas, with the price for gas delivered in a given month equal to the average price of the 'month-ahead' contract in the preceding month at the relevant hub.[22] Suppose that on 1 April 2020 it signs a downstream contract to supply gas in May 2020 to a customer at a fixed price. Without hedging, it is taking the risk that its revenue from the customer is fixed but its cost of gas (under the long-term contact) is uncertain, because costs are indexed to the month-ahead contract whose price will vary over the rest of April. To remove that risk, the buyer could (if trading is sufficiently liquid to make this possible and affordable) make sales and purchases at the hub that will remove the uncertainty and leave it with a fixed margin (positive or negative) on the downstream sale.[23]

The Agency for the Cooperation of Energy Regulators ('ACER') publishes an annual report analysing the functioning of European gas hubs. The most recent (as of writing) report analysed market liquidity in 2017.[24] In terms of volumes traded, it identifies the TTF in the Netherlands and the NBP in the UK as by far the most liquid hubs. Other hubs like the NCG and GPL in Germany or the PSV in Italy trade volumes 10 times lower than at the NBP and TTF. The NBP, TTF,

22 For example, suppose the gas is to be delivered in May 2020. During April 2020 there will be hub trading of a standardised contract for the delivery of gas in May 2020, so on each (trading) day in April 2020 that contract has an observable price. The long-term contract will then determine the price of gas delivered under the long-term contract as the average of those April prices.

23 Specifically, the buyer would: (i) over each of the 22 trading days in April 2020, sell month-ahead contracts with a daily volume equal to 1/22nd of the volume it has sold to its downstream customer; and (ii) on 1 April, buy a month-ahead contract with volume equal to the volume it has sold to its downstream customer. The first of these operations will have revenues that match the costs the buyer incurs under the long-term contract (ie, they cancel out any uncertainty on that cost). The second will give a fixed cost that matches the fixed revenue from the customer.

24 *Results of Monitoring the Internal Electricity and Natural Gas Markets in 2017, Gas Wholesale Markets Volume*, ACER, October 2018, pp30–35.

NCG, GPL and PSV hubs have similar market depth when it comes to spot and prompt hub products but lower when it comes to longer term products.[25] It is generally the case that, except for the TTF and NBP hubs, liquidity decreases substantially when it comes to forward hub products beyond one monthahead of delivery.

In conclusion, we note that prices at hubs in Northwest Europe (notably NBP, TTF, NCG and GPL) are very close to each other, and highly correlated, because of the high level of interconnection within the region. Within that set of hubs, the choice of which hub to use for indexation is therefore unlikely to have a significant impact on the price development of a long-term contract.

4.2 Further considerations on the implementation of hub indexation

Finally, there are other matters that buyers and sellers will need to decide when implementing a hub-indexation formula in a long-term contract. These include: (i) which hub product or products to use (eg, day-ahead, month-ahead, a basket of contracts, etc); (ii) whether to index the price to bid or to ask quotations; and (iii) what information source is used as a reference for the prices of the hub product considered.

The first of these is a relatively complex issue, because of the relationship with flexibility explained earlier, and in the interest of brevity we do not discuss further here, other than to note that in our experience the most common (though by no means universal) approach is indexation to a month-ahead contract.

The latter two issues are, however, unlikely to have a material impact on the revised contract price. Unless there is reason to expect that the bid-ask spread is likely to change in the future, indexation to a bid quotation or to an ask quotation will produce the same contract price (because the month-to-month change in the bid price will equal the month-to-month change in ask price).[26] If, though, there are concerns about the future evolution of the bid-ask spread, a typical solution will be to index the contract price to the average of the two quotations.

In terms of sources of information, there are several entities that publish reference hub prices for OTC transactions, such as ICIS, Argus and Platts. In our experience, they all publish very similar prices.

5. Conclusions

In this chapter we have discussed some of the main challenges and sources of dispute that arise when long-term gas supply contracts have to be adapted via price review to reflect hub prices. We have distinguished between issues that

25 Spot and prompt products refers to contracts for delivery in the very short-term, eg, within the day or day-ahead delivery (spot) or month-ahead (prompt).

affect the price level and issues that affect the choice of indexation in hub-indexed price formulas.

In our experience, price levels are determined by the contractually specified price review criteria. In addition, buyers tend to argue that they have an explicit or implicit right to be able to resell the gas at a profit, and that to do so requires a contract sales price below hub. Sellers, on the contrary, tend to argue that a discount to hub prices would alter significantly the allocation of risks in the contract. If buyers obtain gas at a discount to hub prices, the contract will provide them with a guaranteed risk-free margin. They also argue (correctly, as we explain in detail above) that the flexibility in long-term contracts has value that (all else being equal) justifies a premium over hub price. The discussion around price levels relative to hub prices and how this translates into a margin for the buyer is still a major subject in dispute and very likely the main driver of price reviews in hub-indexed contracts.

With respect to the indexation structure, we have found that parties may disagree about which hubs should be used as a reference, as well as about other issues such as which hub product(s) to index to. In our experience buyers tend to argue that the right hub reference is the hub price corresponding to the market in which the buyer is active, while sellers may argue for hub references that reflect better their opportunity costs and that are less open to influence by the buyer's activities.

In summary, hub indexation has not yet stopped price reviews and price disputes in long-term gas supply contracts. For the time being, we continue to see price reviews and the same core issues of price levels and indexation discussed in the context of price reviews.

The authors thank Anna Masser, Partner at Jones Day, and Agnieszka Ason, Visiting Research Fellow at the Oxford Institute for Energy Studies, for their valuable contributions in the preparation of this chapter.

The potential for price disputes in hub-indexed gas contracts

Valentina Bonetti
Dan Harris
Carlos Lapuerta
The Brattle Group

1. Introduction

Over the last ten years, changes in pricing mechanisms have been a key driver behind price reviews in long-term gas contracts. The financial crisis led to a significant excess of gas supplies, which supported the emergence of European gas trading hubs. Greater liquidity led to price signals that reliably indicated the value of gas. The resulting hub prices were relatively low given the glut of supplies. Although the financial crisis also prompted lower oil prices, they rebounded relatively quickly, so gas and oil prices 'decoupled' in Europe. Oil-indexed gas prices under long-term contracts, which had been the staple of the European gas industry, no longer reflected the value of gas on the market. Decoupling created pressure for 'midstreamers', the large European companies that import gas under long-term contracts for resale to end-users. Established midstreamers found that their customers and competitors could purchase gas directly on hubs at significantly lower prices than under long-term oil-indexed contracts.[1] Midstreamers therefore sought to replace the oil-indexed price formulae with formulae that directly referenced spot or hub prices.[2] Sellers resisted, so many price review negotiations ended up in arbitration. Our experience is that tribunals largely sided with midstreamers, replacing oil-indexed price formulae with formulae referencing hub prices.

When midstreamers first made requests for hub indexation, spot prices were significantly below prices in oil-indexed contracts. Hence, many sellers saw the midstreamers' requests for hub indexation as opportunistic. However, oil prices fell significantly in the last quarter of 2014 and the first quarter of 2015,

1 For a full description of the effect of the financial crisis on the European gas industry and pricing See Anthony J Melling, *Natural Gas Pricing and its Future. Europe as the Battleground* (Washington DC, Carnegie Endowment for International Peace, 2010).

2 Technically, the term 'spot prices' refers to a limited set of contracts for gas delivery in the next day or few days, as opposed to contracts for the delivery of gas in the following month, quarter or year. However, the term 'spot price' is often used to indicate the price of gas at a traded hub, and we adopt this convention in this report.

introducing sustained periods where hub prices exceeded traditional oil-indexed prices. Industry observers and politicians could no longer equate hub prices to low prices, and oil indexation to high prices. Nevertheless, during these periods we still advised midstreamers to request hub price indexation. A hub-indexed price was no longer the 'low price', but still the 'right price'. While midstreamers paid more in the short term, hub price indexation reduced risks sufficiently to make the cost worthwhile.

In Western Europe, the process of replacing oil-indexed price formulae with hub indexation is largely complete. According to the 2018 International Gas Union (IGU) price survey, 70% of gas consumed in Europe is sold based on hub prices.[3]

This chapter explores whether the large-scale switch to hub price indexation signals the end of disputes. After agreeing to hub pricing, are there any remaining issues of friction between sellers and midstreamers?

Hub price indexation has certainly reduced the scope for disputes, by addressing the significant mismatches that arose between the value of gas in the market and the prices in long-term contracts. Nevertheless, three main sources of conflict remain even after the parties adopt hub price indexation:

- Transportation – Most long-term contracts have delivery points that are not located directly at hubs, so the buyer must incur transportation costs to get to a hub. Even with 100% hub price indexation, debates can emerge over the costs of transporting gas to and from the hub.
- Market risk and margins – The transition to hub-indexed prices has changed the level of risk facing midstreamers. Since a midstreamer's margin should relate to the risk it bears, the shift in risk inevitably leads to a debate over the reasonable margin.
- The risk of market manipulation – When choosing a hub, price reviews to date have focused on measures of liquidity such as churn rates, but could benefit from additional analyses into manipulation possibilities, which present a distinct issue. One question is whether an episode of manipulation would constitute a "change in circumstances" that could justify a price review. Cases of manipulation can cause significant damage to either the buyer or seller, but are seldom 'enduring', which is a common condition for price review. It remains unclear how price reviews would address manipulation.

2. Transport costs

Most long-term gas contracts in Western Europe have now moved to 100% hub

3 IGU, *Wholesale Price Survey 2018 Edition, A global review of price formation mechanisms 2005 to 2017* (June 2018), p13: "In Europe gas on gas competition has risen to 70% of total consumption, with oil price escalation at less than half this level."

indexation, but disputes can still arise over the price level. The commercial viability of a European midstreamer depends on a contract price that permits the profitable resale of gas, after incurring costs to serve the end users. Even under hub-indexed contracts, midstreamers have incentives to review the price if it does not allow them to 'market economically' the gas under the contract.

Most long-term contracts have delivery points located at the border of the country where the midstreamer intends to sell the gas. Delivery at the border made sense prior to the liberalisation of the European gas industry, as the seller did not have the right to use the pipeline network within a specific country. With liberalisation, hubs have generally emerged within specific countries, at a distance from the traditional delivery points. To transport the gas to a hub, the midstreamer must typically incur the cost of entering the country's gas transport system.[4]

However, a typical price review clause will only allow the parties to adjust the price, not the delivery point. Hence, even if midstreamers and sellers agree that hub prices represent the best proxy for the wholesale value of gas, there can be disagreement on whether, and how, to adjust the hub price for transport costs from the delivery point. In large contracts, the debate can often have an impact of many millions of euros per year.[5]

2.1 Deduction of transport costs

Historically, to determine the price of long-term contracts at the border, the parties would apply the 'netback' principle[6] illustrated in Figure 1.

The black dots illustrate the points where natural gas prices for customers are available, which is the starting point of the netback analysis. One would then deduct the non-gas costs incurred by the midstreamer between the delivery point – illustrated as grey dots – and the customer. Figure 1 shows grey arrows, which represent the costs incurred in going from the delivery point to the main gas transport network – and black arrows which represent the cost incurred between the main gas transport network and the customer. These costs generally include transportation, storage and distribution costs, commercialisation costs, and a reasonable margin.

The netback pricing methodology was the primary technique for establishing a gas price at the delivery point that allowed the buyer to market

---

4 A notable exception is the Belgian hub of Zeebrugge, which is located outside the Belgian border.

5 Regulated entry costs in Europe vary in the range of €0.1/MWh to €0.7/MWh. Assuming annual volumes of 40TWh under a typical long-term contract implies annual figures of €4 to €28 million per year.

6 See, eg, M Levy, *Gas Price Arbitrations: A Practical Handbook* (London, Globe Law and Business, 2014) 18: "so-called 'netback', where the [non-gas] costs are deducted from the end-user prices [...]."; Anthony J Melling, *Natural Gas Pricing and Its Future: Europe as the Battleground* (Washington, Carnegie Endowment for International Peace, 2010) 16–17: "In order to generate maximum revenue for the state, the 'market-value' or netback value principle was introduced as the basis for natural gas marketing [...] Netback pricing begins with the 'market value' of natural gas in inter-fuel competition (in each market sector) and deducts the costs of transport services and overheads and profit margin to arrive at the 'netback value' at the point of sale."

Figure 1. Netback principle

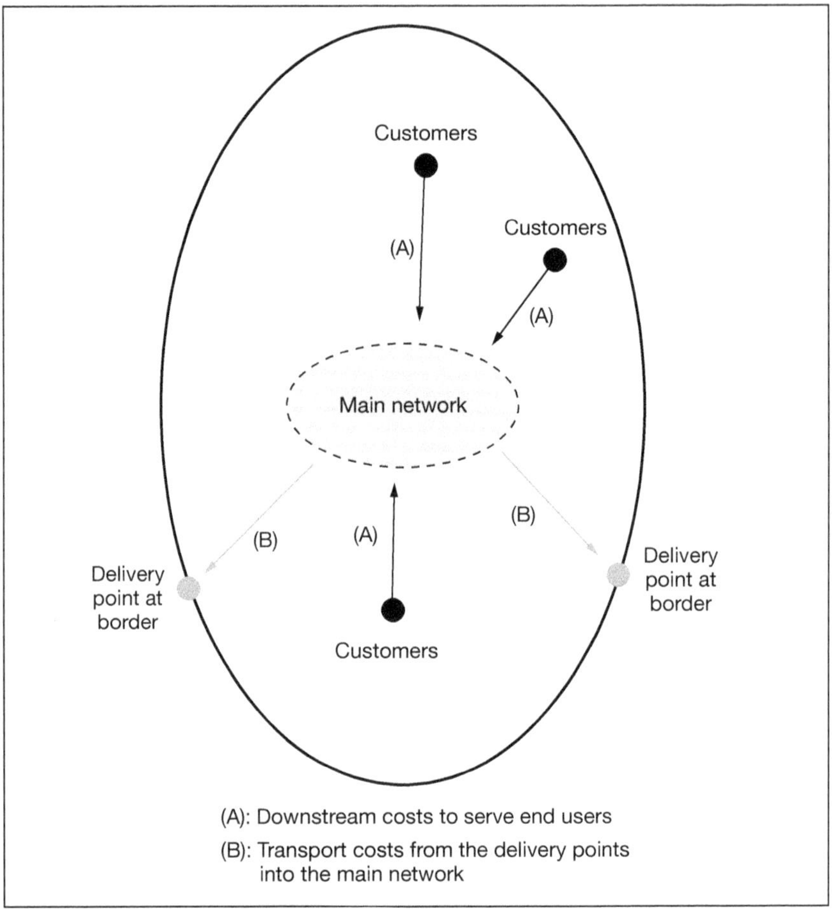

the gas and make a reasonable margin. The seller bore the risk that this price could be more or less than the seller's cost of buying or producing gas. In return, the buyer would commit to buy a minimum volume of gas under a long-term 'take-or-pay' commitment.[7] The parties generally relied on the gas prices customers paid, because they were the only reference points for the value of gas in the market of the buyer.

In practice, the change in perspective from end-user prices to 'wholesale' hub prices can give rise to disputes. Buyers argue that hub prices do not

7 J Stern and H Rogers, *The Dynamics of a Liberalised European Gas Market: Key determinants of hub prices, and roles and risks of major players* (The Oxford Institute for Energy Studies, 2014) 2, NG-94: "In the Continental European context this was addressed by a long term gas contract in which pricing terms were related to oil products with periodic (usually three year) price reviews (or 're-openers') and an Annual Contract Quantity (ACQ) with a minimum Take or Pay (TOP) level set typically at 80–90% of the ACQ level. In this way the 'Seller took the Price Risk' and the 'Buyer took the Volume Risk'."

invalidate netback pricing: one should still subtract transport costs to maintain the contract's allocation of price risk. Assuming that hub prices are reported at the Main Network in Figure 1, the parties can 'skip' the costs associated with the black arrows. They need only consider the costs, and required margin, between the delivery point and the hub, represented by the grey arrows.

Some gas market experts have published articles explaining that transportation deductions remain standard in the industry. For example, Stern and Rogers, from the Oxford Institute for Energy Studies, state that: "mid-stream gas traders [...] can only sell at hub prices, [so] they need to purchase gas below those prices in order to maintain profitability after they have paid transportation [...] and administrative charges related to bringing the gas to the hub."[8]

However, the contractual language often does not mention the netback principle explicitly. In the absence of specific language, we have seen sellers dispute the netback principle, arguing that the parties reached an implicit agreement not to consider transportation costs within the price review clause. The midstreamers respond by arguing that the original contract prices gave them sufficient scope to recover their transportation costs, so the parties in effect adopted a netback approach.

We see two issues that can introduce confusion in the discussion. First is the absence of clear contractual language, and the natural scope for debate over the parties' original intent. Secondly, while agreeing to a price that was indeed consistent with a netback approach, it would have been logical for the seller to resist a price review motivated by changes in downstream costs that were within the control of the buyer or its government. However, that strikes us as a different issue. If one changes the entire pricing mechanism to abandon oil price indexation, and measures the value of gas by reference to a hub that lies some distance from the delivery point, then it is only natural to adjust the hub price reference to measure the value of the gas accurately at the delivery point. A transportation adjustment makes sense, whether or not the transportation costs have changed since the time of contract signature, regardless of who has controlled any subsequent change in transportation costs, regardless of risk allocation issues. Rather, the transportation adjustment becomes necessary simply to provide a 'like-for-like' comparison of the gas at the delivery point to the gas sold at a hub, in a different location.

2.2 Value of transport costs

Buyer and sellers might also agree on the principle of deducting transport costs, but disagree on the value of such costs. In particular, buyers tend to consider transport costs based on regulated network tariffs, which are generally publicly

---

8 J Stern and H Rogers, *The Dynamics of a Liberalised European Gas Market: Key determinants of hub prices, and roles and risks of major players* (The Oxford Institute for Energy Studies, 2014) 66, NG-94.

available. However, sellers often seize upon the significant difference between the published regulated tariff, and the value of gas transport capacity in the secondary market. In recent years, European gas markets have generally experienced a surplus of transport capacity. Causes included the lower demand following the Global Financial Crisis, and security of supply regulations, which at times oblige gas system operators to build more capacity than the market would demand for purely commercial purposes. As a result, the market value of transport capacity on many routes has fallen below the cost indicated by regulated tariffs.

The development of liquid hubs in Europe has made it easier to measure the market value of the transport network, and has exposed the difference with regulated tariffs. The difference between hub prices in two countries should reflect the market value of transporting gas from one country to the other. Instead of transporting gas from hub A to hub B, a gas company can sell its supplies at hub A, and purchase an equivalent amount at hub B. Actual transportation makes no sense if it costs more than the strategy of selling at one hub and purchasing at another. Supply and demand conditions can lead to similar prices at hub A and hub B, which suggests little value to transporting gas between the two hubs.

Figure 2 illustrates with an example, where Hub_B is the local hub in the country of the buyer, and Hub_A is the hub in a neighbouring country where the seller produces gas. The price difference between the two hubs is 2, calculated as 10 minus 8. On the other hand, regulated tariffs may suggest transport costs of 5, calculated as the sum of the 2 to exit country A, and the tariff of 3 to enter country B.

In this example, a buyer would argue it can only compete in its market if the price at the delivery point on the border is Hub_B - T_2 or 7. The buyer's competitors would purchase gas at the local hub for a price equal to Hub_B, or 10. If the buyer paid a price of 7 for the gas at the border, and incurred transport entry cost of 3, it would pay 10 overall, or the same as its competitors at the hub.

However, the seller may resist a deduction of a transportation cost of 3 from the price at the destination hub. The seller may argue that it would end up receiving a price of 7, which is less than the 8 that it could obtain by selling locally at Hub_A, and without incurring the additional transportation costs to deliver the gas to the midstreamer.

In this example one could say that the transportation costs to Hub_B are stranded, in the sense that the regulated tariffs for transportation exceed the value that one can add to the gas by transporting it. The efficient solution might actually be to sell the gas locally at Hub_A, as opposed to transporting it to Hub_B, but the long-term contract requires the delivery of gas to the country of the midstreamer. The long-term contract may well have been the efficient outcome under different market conditions that prevailed at the time of its signature. If a subsequent stranding has occurred, the question is who bears the costs? Deducting the regulated tariff in the example above allocates the stranding risk

Figure 2. Transportation costs and price differential

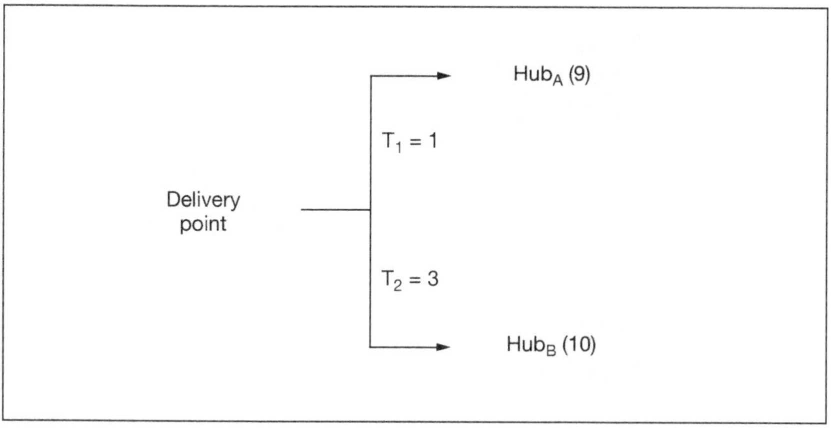

to the seller. Giving the seller the price at Hub$_A$ would force the midstreamer to bear the consequences of stranding. The answer depends on an inquiry into risk allocation under the contract.

2.3 Other related disputes

Similar disputes can arise with respect to other infrastructure like storage and regasification terminals. At times, a midstreamer signs a long-term contract committing it to a relatively fixed price for either storage capacity or regasification capacity. The long-term infrastructure contract may in fact complement a long-term gas supply contract. The market price of the infrastructure may have fallen below the price that the midstreamer has committed to pay over the long term, in some cases very significantly. A separate class of disputes involves midstreamers asking for price reductions to recover losses associated with infrastructure.

As above, the midstreamer has an incentive to argue for a deduction from the hub price to reflect its payments for complementary infrastructure like storage or regasification. Again, the seller may resist bearing the risk and costs of the stranded infrastructure. The seller might at most accept a deduction reflecting the far lower market value of the associated infrastructure.

Regasification presents a unique issue because several terminals in Europe enjoy exemptions from third-party access requirements. They may not publish a regulated tariff, or they may publish a tariff that applies to certain volumes not booked over the long term at the terminal. The actual tariff applicable to long-term deliveries could differ significantly from the published tariff. In this case a transparency issue can compound the issues associated with potential stranding of the midstreamer's costs.

Unfortunately, full disclosure of the midstreamer's regasification costs may not resolve the conflict with the seller. A seller can object that an obligation to underwrite the midstreamer's regasification costs, by absorbing a deduction from the hub price, will eliminate the midstreamer's incentive to minimise the costs of the regasification service. One solution could be to set a cap on allowed regasification costs based on benchmarking the cost of other LNG terminals, but then the dispute devolves into the appropriate benchmark, and potential adjustments to the benchmark.

3. Market risk and margins

Other kinds of disputes relate to disagreements over the reduction of risk introduced by the switch to hub indexation, and the implication of reduced risk for the midstreamer's margin.

Producers might argue that midstreamers who resell in a liquid market now have less exposure to volume risk, since they have an outlet for excess supplies, and can easily buy gas in the event of a shortfall. Sellers might also argue that the transition to hub-based pricing has significantly reduced pricing risks. Hub price indexation can allow midstreamers to purchase gas under long-term contracts at a comparable price to their competitors who access gas directly at the hub. Moreover, hub prices now set end-user prices in much of Europe, so hub price indexation aligns a midstreamer's gas purchase costs with its gas sales prices. Sellers often argue that midstreamers now deserve lower margins to accompany their lower risks.

Hub indexation has indeed reduced risk substantially for midstreamers relative to the turbulent years of 2009–13, when the decoupling of oil and gas prices threatened the survival of several large importers. However, perceptions of risk reduction depend on the reference point. Many midstreamers can point to an earlier reference point: the time of contract signature. Midstreamers and sellers signed most existing long-term contracts when natural gas purchases and sales all occurred at oil-indexed prices. Oil-price indexation historically reduced the risks facing midstreamers, as they simultaneously resold gas to end-users at oil-indexed prices. Many observers view the original goal of long-term contracts as minimising the midstreamer's price risk, to enable it to bear the volume risk associated with a take-or-pay commitment.[9]

Adopting an earlier reference point, one could describe the recent shift to hub indexation as simply restoring a historical low-risk paradigm for the midstreamer. Adopting a broader time horizon, high risks were largely an anomaly between 2009 and the switch to hub price indexation. If the buyer is simply returning to the same type of risks borne earlier, it may argue that it deserves to retain the margin that it earned earlier, as opposed having the margin reduced.

We have seen some experts use the dismissive term 'monopoly margins' to

---

9 See fn 7 above.

describe the margins earned by midstreamers prior to the financial crisis. Years ago midstreamers earned higher margins, and they now earn lower margins. We see no basis to conclude that the more recent profitability of midstreamers indicates an appropriate margin, particularly in countries where some midstreamers are exiting the business or curtailing operations due to low profitability. We recommend engaging with the question why margins were higher previously. If risks were indeed low at the time, yet the midstreamer stood to earn a relatively high margin, it is reasonable to inquire whether the midstreamer's long-term contract reflected an implicit agreement to protect the margin from subsequent erosion. In our experience, the answer tends to vary from contract to contract. Some contracts have pricing mechanisms and price review clauses that seem designed to preserve margins; others appeal to notions like a 'competitive price' or the prevailing prices in other long-term contracts, which are inconsistent with margin preservation.

We now turn to a separate argument, that hub price indexation eliminates virtually all price and volume risk. The simplistic argument is that, if the midstreamer cannot find a customer, it can nevertheless dump the gas at the hub. In such a case, the midstreamer would receive a hub price for dumping the excess gas supply, which would presumably match the price that it paid the seller. The midstreamer would presumably incur no loss. A seller might argue that the midstreamer paying a hub price faces no downside, and only upside, so there is no material risk and no need to worry about the midstreamer's margin.

Midstreamers can legitimately respond that liberalisation has introduced other risks, such as the development of intense competition. Midstreamers have invested in the development of large sales forces and supporting business infrastructure. Competition now exposes midstreamers to a loss of market share, to the possible need to dismiss staff, and in extreme cases to the bankruptcy of the entire organisation. Midstreamers can argue that hub-indexed contracts reduce risk somewhat, but do not yet restore them to the low-risk position that they occupied when initially signing their long-term contracts.

Often the hub-indexation in long-term contracts tends to be to short-term hub products, while downstream contracts tend to be for fixed prices, or indexed to longer-term hub products. Midstreamers implement 'hedging' strategies[10] to reduce as much as possible their price risk, but hedging cannot eliminate the risk fully.

---

10 Hedging is a standard practice among midstreamers, consisting of supplementary purchases and sales of gas at hubs, undertaken on a routine basis whenever they sign end-user contracts. The hedging transactions involve the same gas volumes, so the midstreamer does not actually buy any more gas than it sells. However, the midstreamer would sell gas at a fluctuating spot price, while purchasing gas at a fixed price. The supplementary hedging activity reduces risk, locking in the margin of the relevant downstream transactions. In the absence of hedging, the midstreamer would purchase gas at a fluctuating price, as typically dictated by the spot price indexation in its long-term gas purchase contracts, but it would resell the gas for the fixed price demanded by the customer. It is risky to purchase at a fluctuating price for resale at a fixed price.

The UK provides a good example of the continuing risks that companies face after the transition to hub price indexation. UK midstreamers purchase gas at spot market prices, with deductions for transportation if any of that gas is delivered outside the UK pipeline network. Yet whether they make or lose money depends on fluctuating market conditions. Their fluctuating financial performance derives from the commitment to serve relatively small end users, as we explain further below.

The UK regulator has assessed the profitability of midstreamers, and has concluded that they can lose money at times when hub prices rise rapidly.[11] The regulator cited the "cost to suppliers of changing their retail prices, which means that the retail price will not always reflect the wholesale cost that the supplier is facing at a particular point in time".[12] That is, when spot prices increase rapidly, midstreamers hesitate to pass through all cost increases immediately to end users. Large end-users tend to be sophisticated, and can understand the volatility of commodities markets, but small end-users react with suspicion to large, unexpected price increases. Midstreamers therefore try to keep the prices on offer stable, as opposed to updating them each day in line with daily changes in commodity prices. Maintaining stability avoids a loss of confidence by smaller end-users, but introduces risk to the midstreamer.

As hub prices rise, buyers will eventually pass through the increases by insisting on higher prices in any new contract signed by a smaller end-user. However, the midstreamer will tolerate losses until naturally replenishing the customer portfolio with new, updated contracts. Midstreamers hope that the opposite will occur – if wholesale prices drop unexpectedly, perhaps they can postpone passing through the reductions to end users, and make money on the decline. However, there are no guarantees. It is reasonable to conclude that the pricing dynamics described above introduce risk: losing money when spot prices spike, in exchange for the hope of offsetting the losses with a transitory windfall as spot prices fall.

A question naturally arises whether the 'in any case' or 'market economically' clause should allow a midstreamer to claim a 'below hub' price to protect its margin, at times when high spot prices might impose losses on it or its competitors. Sellers might claim that the parties' switch to hub price indexation justifies the abandonment of a focus on the shorter-term profitability of resale in the retail market, in favour of a more long-term view. With a long-term perspective, we see the logic in trusting that the payment of a hub-reflective price should permit the midstreamer to earn a reasonable margin over the long run. As indicated above, over the long run one might expect to offset losses during wholesale price spikes, with transitory windfalls from sudden dips in wholesale prices.

11 Ofgem, "Energy Supply Probe – Initial Findings Report", 6 October 2008.
12 Ofgem, "Energy Supply Probe – Initial Findings Report", 6 October 2008, p74.

However, other cases raise broader issues than simple spikes or dips in wholesale prices. The midstreamer may be able to demonstrate a generally depressed market, in which other midstreamers are actually exiting the business or curtailing operations. If the problem relates to intense competition, and appears to be of a structural nature, we see no basis to dismiss such a situation as simply the 'swings and roundabouts' associated with spikes or dips in wholesale prices. Rather, a depressed market may well extend until the next price review period. If so, the midstreamer may have a valid basis to invoke the market economically clause to justify a discount relative to a hub price.

4. The risk of price manipulation

In this final part of the chapter, we discuss how indexation to hub prices can introduce a risk of price manipulation, what parties can do to minimise the risk, and whether an increased risk of manipulation could give rise to a price review.

While oil prices no longer reflected the market price of gas in Europe after 2009, buyers and sellers could at least trust that the oil price was not vulnerable to manipulation. The oil market was sufficiently deep and liquid that neither the buyer nor the seller could have any influence over the oil price. This was one of the reasons why parties in long-term gas contracts referred to crude oil prices, or to the prices of refined oil products.

However, products traded on gas hubs, and to which buyers and sellers may wish to index gas prices, are typically less liquid than products traded on international oil markets. Liquidity and the risk of price manipulation are closely related: it is difficult, or at least prohibitively risky and expensive, to manipulate prices on a liquid hub. To see why, consider the following example, which is central to the discussion that follows.

Suppose that a buyer and seller index the price in a large gas contract to the average price of month-ahead ('MA') products on the last trading day of the month before delivery. So for example, the price of MA contracts on Thursday 31 January determines the long-term contract price for February. Typically, the buyer takes 3,000 GWh per month.[13] Suppose that the market for the month-ahead product is relatively illiquid. The buyer decides to depress the month-ahead price, by selling month-ahead gas products on 31 January at below the prevailing bid price. Specifically, suppose that the buyer sells 100 GWh of gas at 0.5 €/MWh below the bid price, and as a result manages to decrease the average month-ahead price by 0.1 €/MWh. Clearly, this action will cost the buyer money, because it is obtaining a lower price than it could have got for its gas. The buyer has lost 0.5 €/MWh x 100 GWh, or €50,000. However, the buyer has made a much larger gain, because it is able to buy 3,000 GWh under the long-

13 This is equivalent to a contract with an annual delivery volume of over 3 bcm per year. These kinds of contract volumes are very typical in the European gas industry.

term contract at a price that is 0.1 €/MWh lower than otherwise. Accordingly, the buyer's manipulation has saved it €300,000. The net benefit of the manipulation to the buyer is €250,000. We could imagine a similar manipulative scheme by a seller, who buys gas at above the market price, inflating the prices received for sales of gas under long-term contracts.

The example is not only hypothetical – the US has witnessed several cases where traders have manipulated gas prices as described in the example above. For example, in 2007, the FERC accused Energy Transfer Partners, LP (ETP) of manipulating natural gas prices at the Houston Ship Channel (HSC) trading hub from December 2003 to December 2005. Specifically, the allegation was that ETP deliberately lost money on its index-setting sales of monthly fixed-price natural gas at HSC, to benefit its physical and financial positions that benefited from decreases in the published HSC index price. The case settled for $30 million.[14] Part of the issue in that case was that traders had grown so used to simply indexing contracts to the hub price, that actual fixed-price sales, which underpin the derivation of the hub price, had become relatively few in number. ETP was therefore able to depress the index price at relatively low cost.

The examples above illustrate four points. First, liquidity and the reliability of hub prices – so their robustness against manipulation – are closely related. If, in the example, above, the market for the month-ahead product was very deep and liquid, the manipulator would need to sell a much larger volume of gas to move the market price. Selling a larger volume of gas at a loss increases the cost of the manipulation. At a certain point, the scheme described above is no longer profitable. Hence, the price is robust against this kind of manipulation attempt. Of course, in the example above the contract price was particularly vulnerable to manipulation because the parties had chosen to fix the price using the average trading results of a single day. Indexing the contract price to the average price over the entire month would make it more difficult and costly to manipulate the price.

Secondly, the incentive to manipulate depends on the size of the gas contract. The example above is roughly based on a 3 bcm/year contract. If the contract is for 9 bcm/year, the buyer's (or the seller's) profits from the manipulation would be three times higher, incentivising either party to tolerate higher losses from short-term trades in an attempt to move the market price. In our experience, market participants do not always appreciate that indexing large gas contracts to a hub can create the incentive to manipulate. We have heard market participants dismiss the risk of price manipulation of a particular hub on the basis that "it has never been manipulated before". The absence of

14 For details of this and other cases see Shaun D Ledgerwood, and Paul Carpenter, "A Framework for Analyzing Market Manipulation" (2012) in *The Review of Law & Economics* (DeGruyter/Berkeley Electronic Press, forthcoming). Available at SSRN: https://ssrn.com/abstract=1811764 or http://dx.doi.org/10.2139/ssrn.1811764.

prior manipulation may reflect nothing more than the lack of incentives prior to the widespread switch to hub-price indexation. The temptation may present itself now that many billion cubic meters of gas are priced at hub.

Thirdly, our experience has been that discussions of hub pricing focus too narrowly on the assessment of the liquidity of the hub as a whole, as opposed to specific products. Rather, the discussion should consider whether the specific underlying contracts or products are liquid or not liquid. Accordingly, any analysis of liquidity should focus on specific hub products chosen for indexation purposes: day-ahead, week ahead and so on. A particular hub may have a very liquid market for day-ahead products, but an illiquid market for year-ahead products. It is not informative to call the hub liquid or illiquid – rather, in such a case parties could safely index to day-ahead prices on the hub, but may validly refrain from indexing prices to year-ahead products. Similarly, considerations of liquidity often focus on measures that are relatively crude, and therefore do not sufficiently address the risk of market manipulation. The 'churn rate' is a measure of liquidity that market commentators cite frequently. In effect the churn rate measures average traded volumes across all products – a hub could have a churn rate that seems to indicate liquidity, but with some illiquid products vulnerable to manipulation.

Fourthly, the Houston Ship Channel case shows that hub-price indexation can have an unanticipated consequence. Hub-price indexation had become so common that relatively few market participants actually bought and sold gas at fixed prices, which were the transactions necessary to determine the index each day. Many participants in effect sat on the sidelines, trusting that others would set a reasonable price. The prevalence of hub-price indexation ended up reducing the liquidity of the trading that set the hub price.

To minimise the risk of manipulation, the parties should carefully analyse the liquidity of the products to which they are considering indexing the long-term contract, and specifically evaluate whether, given the contract sales volume, it could be profitable to engage in the type of manipulation that we describe above.

Having set out the risk of manipulation, and the analysis that diligent parties should undertake to minimise the risk, we now turn to whether market manipulation could give rise to future price reviews in long-term contracts. Disputes could arise if the parties have indexed their long-term contract to a product without sufficient analysis, and realise later that the product is illiquid and vulnerable to manipulation. Alternatively, the product may have been liquid when the parties agreed the formula, but since became illiquid. Either way, the question arises whether an escalated risk of manipulation could justify a price review under many existing review clauses. The issue is not clear cut. On the one hand, an increased risk of manipulation would constitute a 'change in circumstances' under many price review clauses. On the other hand, it is not clear how an increased risk of manipulation "affects the value of gas". Manipulation does not cause the value of gas to change. Rather, it leads the contract formula to misstate the value of gas. A related question is whether the increased risk of manipulation is enduring.

Assessing the risk of manipulation will involve complex analysis and the selection of parameters on which reasonable people can disagree. Part of the analysis may be econometric in nature. Moreover, the results could change over time. Hence, it may be an analytical challenge for a claimant to establish that the risk of manipulation is enduring.

In principle, the vulnerability of an index to manipulation would motivate both parties to adjust the formula. As the example above illustrates, either the buyer or the seller could be the victim of a manipulation. However, disputes could still arise if one party disagrees with the other's assessment of the manipulation risk. Alternatively, both parties may agree on the problem, but not on the solutions, or one party may stand to lose or gain more than the other from a proposed change to the formula.

Ultimately, the typical European price review clause could be too blunt an instrument to deal with the issues described above. Legal specialists may see an avenue to address the risk of market manipulation under the typical 'change of index' clause, which many long-term contracts contain. These clauses typically allow the parties to substitute a price reference, if for example it is no longer published, or the product is no longer traded. A notable recent case was High Sulphur Fuel Oil (HSFO). Many long-term contract price formulae referred to the price of HSFO, but the index was phased out due to environmental regulations. As a result, buyers and sellers needed to agree to refer to a different fuel, usually Low Sulphur Fuel Oil. It would be prudent for buyers and sellers, when inserting clauses referring to hub prices in price formulae, to modify the 'change of index' clause as well. If a product becomes, or is revealed to be, illiquid and hence vulnerable to manipulation, specific language could permit the parties to change the price formula without having to meet the typical criteria of a price review clause.

5. Conclusions

The convergence to hub pricing in European gas markets has largely resolved one of the main sources of gas contract disputes over the last ten years: the mismatch between long-term contract prices, prices in liquid wholesale markets and prices in end-user markets. However, the potential for disputes remains. First, hub-indexation does not necessarily imply a price level that satisfies both parties. Conflicts over the costs of transportation, storage and other supporting infrastructure can still arise. Furthermore, accepting the principle of hub price indexation does not end the debate over a reasonable margin. Rather, changed market conditions can intensify that debate. Finally, the risk of price manipulation raises an issue concerning the reliability of the specific hub products selected for indexation, and another issue concerning the future grounds for revising the contract price in the event of actual manipulation.

Some may believe that the remaining disputes are likely to be of a lower magnitude than the recent phase of cases over decoupling. However, the recent wave of arbitration proceedings has arguably lowered the point at which buyers and sellers are willing to go to arbitration. Many buyers and sellers shied away from arbitration before 2008, but companies have since learned how to conduct proceedings without disrupting the necessary day-to-day relationship of contract management and implementation. Within Europe at least, there is a greater appreciation of arbitration proceedings as a neutral forum for resolving differences that would otherwise bring management on both sides to an impasse. We therefore expect continued disputes over the next few years, especially given that the new world of hub-price indexation continues to present risks for both sellers and midstreamers.

Gas price reviews in Asia-Pacific

David Jenaway
Allen & Overy

1. Introduction

The Asia-Pacific region is home to the world's largest importers of LNG[1] as well as some of the world's major LNG producers.[2] Notwithstanding, price reviews being resolved by some form of formal dispute resolution have, to date, been few and far between. Indeed, it appears that as at mid-2019, only one dispute relating to the price under a contract for the supply of LNG to Asia-Pacific buyers has been publicly reported as being submitted to international arbitration.

At first glance this may be surprising, especially given the size of the market and the prevalence of long-term contracts in this region. It also presents a contrast with other regions of the world, where arbitration has been more common in this context.

Why has this been the case? What has happened 'in place' of arbitration in Asia-Pacific? What role is arbitration expected to play in the region going forward? These are the primary questions that will be addressed in this chapter. It will be seen that despite its 'slow start', there is reason to expect that international arbitration (and other forms of formal dispute resolution) will feature more heavily in the resolution of gas pricing disputes in Asia-Pacific in the near future.

2. LNG pricing in Asia-Pacific

Gas pricing in the Asia-Pacific exist in a rather unique and specific context. The dynamics in this region are unique, in terms of the natural gas markets, regulatory environments, contracts and dispute resolution preferences. In focusing on "Asia-Pacific", we are primarily examining the Asian countries

1 The world's five largest importing countries are all located in Asia being, in descending order, Japan, China, South Korea, India and Taiwan. China was predicted to overtake Japan to become the largest importer in 2019. Asia is also currently driving international LNG demand growth, led by China, South Korea, India and Pakistan: see International Gas Union, *2019 World LNG Report*, p6, 16.

2 Most notably Australia, which is currently the world's second largest exporter of LNG and predicted to overtake Qatar to become the largest exporter in the course of 2019. Others include Malaysia, Indonesia, Papua New Guinea and Brunei: see ICIS, Global Energy Outlook 2019 at www.icis.com/explore/resources/news/2019/01/10/10304951/lng-outlook-2019-part-one-asia/; International Gas Union, *2019 World LNG Report* p14.

which import LNG. This is a very significant market, currently accounting for over 70% of global LNG demand.[3] Within this set of countries, there are certain commonalities, as well as some important differences, which bear upon their individual price review contexts and experiences.

Japan was the first country in Asia to start importing LNG, in the late 1960s. This was followed by Korea in the mid 1980s and Taiwan in the 1990s.[4] Together, these three countries (often referred to as 'JKT') comprise the "traditional" Asian LNG importers.

In addition, there is a set of relatively new and rapidly growing LNG importing countries in the region, including China (which started importing in the mid-2000s and is now the world's second largest LNG importer), India, Thailand and Pakistan and Bangladesh (which all began importing since 2000).

LNG imports into Asia have traditionally been arranged via long-term contracts (SPAs), many of which have underwritten the development of large-scale LNG projects.[5] The price is commonly determined in accordance with a formula stipulated in the contract, being a linear function of crude oil prices – typically 'JCC', the Japanese customs cleared crude oil price.[6] Specifically, the formula often takes the following form:

$$P_{LNG} = A \times Bn + C$$

where *Bn* is the average of JCC values across a specified period (usually the preceding six to nine months, with a 'lag' of three months).[7]

This pricing method was devised when Japan began importing LNG.[8] At this time there was no 'market' for LNG in Asia (or Japan), in the sense that LNG was not a tradable commodity.[9] Importers – who were typically government-owned entities – would contract for supply of LNG on a long-term basis and sell it directly to consumers or wholesalers, in competition with other fuels, including crude oil.[10] Unlike LNG, oil had reliable, published global benchmarks and provided, at

---

3 International Gas Union, *2019 World LNG Report* p16. See also https://asia.nikkei.com/NAR/Articles/Vivien-Yang-Long-term-LNG-contracts-good-for-Asia. Asia has dominated global LNG trade since the late 1970s (JP Stern and HV Rogers *Challenges to JCC Pricing in Asian LNG Markets* (Oxford, OIES, 2014) 3).

4 A Flower and J Liao, "LNG pricing in Asia" in JP Stern (ed), *The Pricing of Internationally Traded Gas* (Oxford, OUP/OIES, 2012) 338.

5 This is still dominant, although alternative arrangements are increasingly being used too, such as short term and spot contracts.

6 Or the 'Japanese Crude Cocktail'.

7 And *PLNG* is the price of LNG in \$/MMBtu; A and C are constants negotiated by the buyer and seller.

8 JP Stern, "International gas pricing in Europe and Asia: A crisis of fundamentals" (2014) 64 *Energy Policy* 46; JP Stern and HV Rogers *Challenges to JCC Pricing in Asian LNG Markets* (Oxford, OIES, 2014) 17; Flower and Liao (see fn 4), 338, 358.

9 LG Henneberry and GM von Mehren, 'Considerations in natural gas and LNG price review claim quantification' (2015) *Journal of World Energy Law and Business* 2.

10 *ibid*; Flower and Liao (see fn 4); Stern (see fn 4), 4. Thus, there was no 'gas-to-gas' competition: see S Anway and GM von Mehren, "The Evolution of Gas Price Review Arbitrations" in JW Rowley (ed), *The Guide to Energy Arbitrations*, 3rd edn (Global Arbitration Review, 2019). At this time, oil was the main competing fuel to gas for power generation in Japan (Stern and Rogers (see fn 8) 17). This process was highly regulated, and the importers faced little competition (Stern (see fn 4), 4).

this time, the best option for an index to use in LNG pricing.[11] The formula was adopted when Korea and Taiwan began importing LNG, and ultimately across the region more broadly.[12] It is still dominant today in long term SPAs for the sale of LNG into Asia (albeit with some variations, particularly in some of the more emergent markets).[13] In pricing negotiations, attention is usually focused on the 'slope' ('A' in the above formula).[14]

3. The history of price reviews in Asia-Pacific – negotiation, but few 'formal' reviews

It appears that price review clauses only started to be regularly included in long term SPAs for delivery to Asia from the early 1990s.[15] Before this, contracts sometimes included provisions requiring the parties to 'meet and discuss' prices in certain circumstances[16] without mandating a revision or escalating the discussions to formal dispute resolution. This led to uncertainty as to what should happen if the parties could not reach agreement.

Further, as compared to (for example) the European experience, there have also been relatively few formal disputes relating to pricing under Asian LNG contracts. Indeed, according to public information, only one arbitration has been commenced relating to a pricing dispute under a contract for the supply of LNG to the Asia-Pacific region.[17] This was commenced by KOGAS, a buyer from Korea (and currently the second-largest single importer of LNG in the world) against the North West Shelf joint venture, an Australia seller operated by Woodside Petroleum.[18] The dispute reportedly relates to a price renegotiation under a supply contract which terminated in 2016 (and, thus, is retrospective).

Whilst parties to Asian LNG contracts have rarely resorted to arbitration in resolving their pricing disputes, there have certainly been some form of price reviews in this region. These have occurred primarily via negotiation (although there has also been some resolved by other methods, including expert determination). It is difficult to monitor these, as LNG prices and the details of LNG price reviews are not transparent in Asia-Pacific.[19] However, some information has been gathered (from *ad hoc* pieces of disclosed information and a small number of dedicated studies) which enables observations to be made.

---

11 *ibid*. Although markets recognised the 'imperfect' nature of the formula, in terms of using oil as a proxy for tracking the value of natural gas. See also Anway and von Mehren (see fn 10).

12 See Stern and Rogers (see fn 8), 18 and Flower and Liao (see fn 4), 358.

13 Including using alternatives to JCC, such as Brent index. Note that different ways of pricing *short term* cargoes have developed: Flower and Liao (see fn 4), 338.

14 Stern (see fn 8), 46; Flower and Liao (see fn 4), 339. For example, where A = 0.1485, often expressed as a slope of 14.85%.

15 Flower and Liao (see fn 4), 352.

16 A Ason, "Price reviews and arbitrations in Asian LNG markets" (Oxford, OIES, 2019) 1.

17 At least, to the extent of publicly available information.

18 See www.gastoday.com.au/2018/02/13/kogas-and-nws-lng-in-arbitration/. Given the confidential nature of the proceedings, the exact commencement date of the arbitration cannot be ascertained.

19 Stern (see fn 4), 5–6.

For example, a series of negotiations reportedly took place between sellers and Japanese buyers after oil prices collapsed in 1985.[20] They were led by sellers, who were hoping to obtain increased protection at low oil prices. Ultimately this led to the introduction of a variant to the pricing formula known as the 'S-Curve', involving a decrease in the slope of the pricing curve outside an agreed middle-range of JCC values.[21] Buyers were prepared to accept this, as it gave them protection at higher oil prices and, as a result, S-Curves were commonly adopted in most existing Japanese contracts, and new contracts, by the early 1990s.[22]

Another product of these negotiations was the introduction of an 'applicable range' into some contracts, requiring the parties to negotiate if oil prices moved outside a stipulated range.[23] One specific example of a price reviews was the apparently protracted renegotiation between Tokyo Electric Power Company (TEPCO) and Abu Dhabi Gas Liquefaction Company (ADGAS) in the early 1990s, where the price was reportedly reopened several times.[24] As explained below, this criteria of 'applicable range' is not uncommon, but does not appear to have been generally adopted in the SPAs in the region.

From 2003, various isolated negotiations were reportedly triggered contractually – again in contracts with Japanese buyers – when oil prices increased and moved outside the 'applicable range'.[25] Many of these negotiations took place between 2003 and 2010 and were often protracted.[26] Prior to 2010, price review negotiations were primarily limited to contracts with Japanese buyers. There were only isolated examples with buyers from Korea, Taiwan and China.[27]

The first real 'wave' of price reviews across the region took place between 2010 to 2014, when various sellers reportedly requested price reviews from Asian buyers[28] and managed to secure price increases via negotiation.[29] For example, in 2012 the Indonesian government reportedly triggered negotiations

---

20 Flower and Liao (see fn 4), 340.

21 *ibid.*

22 *ibid*, 340–41.

23 *ibid.*

24 Ason (see fn 16), 2; Flower and Liao (see fn 4), 353. These negotiations took six years to resolve.

25 JCC exceeded the upper limit of $29/bbl in many of the Japanese contracts (Flower and Liao (see fn 4), 343).

26 Many of these negotiations would have involved *removing* the S-Curve and reverting to a straight-line model (Flower and Liao, (see fn 4), 343–44). The negotiation conditions were difficult, as buyers and sellers were trying to find pricing that would be acceptable to them, at much higher oil prices than were foreseen in the late 1980s/early 1990s, when S-curves and applicable ranges were introduced.

27 Flower and Liao (see fn 4), 352–53.

28 Across the region, including various buyers from Japan, Korea and China (Ason (see fn 16), 1).

29 Sellers were able do this due to various factors including strong demand growth in Asia, including by reason of Fukushima in 2011: see https://uk.reuters.com/article/uk-lng-prices-analysis/analysis-lng-producers-seize-upper-hand-in-global-contract-reviews-idUKBREA0K0ZV20140121.

30 The contract had been entered in 2002, and was amended in 2006. The amended contract included a price review clause: see www.thejakartapost.com/news/2011/11/24/govt-team-review-lng-sales-china.html; www.spglobal.com/platts/en/market-insights/latest-news/natural-gas/010812-indonesia-to-begin-renegotiating-tangguh-lng-price-with-chinas-cnooc-in-jan.

with CNOOC for sales of LNG from Indonesia's Tangguh Project, concluding in 2014 with a substantial increase to the price,[30] and Petronas reportedly secured a substantial increase in sales to KOGAS from the Tiga project in Malaysia.[31]

From 2015, the dynamic shifted and a series of buyers were reportedly able to secure price decreases via negotiation. An example was Petronet, India's largest buyer of LNG, reportedly obtaining a reduction of almost 50% of the price payable to RasGas, a Qatari seller, in 2015.[32] This result appeared to inspire other buyers. Since this time (for example) Japanese buyers have reportedly triggered price review discussions (or are intending to) with sellers from a range of countries,[33] Petronet also reportedly secured a reduction from ExxonMobil for LNG supplied from the Gorgon project in Australia[34] whilst three major buyers from Korea, Japan and China (KOGAS, JERA and CNOOC) apparently formed an alliance in 2017 that assisted their respective positions.

4. The characteristics of price review clauses in Asia-Pacific

The above gives some context to the introduction of price reviews in Asia-Pacific, and their development. Over time, price review clauses in this region have begun to homogenise, with a number of characteristics, some of which distinguish them from their counterparts in other regions.

First, the clauses typically provide for reviews to occur at, and only at, regular intervals after deliveries commenced.[35] These reviews were generally to be conducted every five years, although sometimes this was extended to ten years.[36] By contrast to other jurisdictions, the clauses generally do not contain an allowance (or a precondition) or certain other preconditions to be met, such as a material change in the market or other circumstance. Rather, the only precondition is that the relevant time period has expired. As the above discussion demonstrates, however, such preconditions can exist; they are just not common.

Secondly, the clauses generally require as a first step a negotiation period, during which the parties are required to attempt to agree on a revised price.[37]

31 This reportedly involved tripling the contract price. KOGAS was also subject to price review negotiations with sellers from Yemen and Russia during this period.

32 Ason (see fn 16), 2; E Van Geuns, "The Dawn of LNG Price Reviews in Asia Pacific" (2016) 56 *Journal of the Australian Petroleum Production & Exploration Association* 1. Reportedly, this contract did not contain a price review clause. In 2015 Petronet failed to lift the amount required under its take or pay obligations, and eventually managed secure this significant price reduction. The revision also reportedly involved a change in the index in the price formula, from JCC to Brent, see https://economictimes.indiatimes.com/industry/energy/oil-gas/qatar-scraps-floor-cap-in-gas-price-formula-with-india/articleshow/50424196.cms.

33 Ason (see fn 16), 2.

34 See www.lngworldnews.com/indias-petronet-inks-revised-deal-with-exxonmobil-buys-more-lng/; www.gasstrategies.com/information-services/lng-business-review/rasgas-petronet-lng-renegotiation-deal-one-year.

35 L Lau "Price reviews in the Asian LNG market" (2014) 32 *International Energy law Review* 170.

36 Ason (see fn 16), 8.

37 Ason (see fn 16), 6–8; Flower and Liao (see fn 4), 353.

Whilst the enforceability of such provisions (colloquially described as having the potential to be 'agreements to agree', depending on the drafting) is outside the scope of this chapter, parties generally adhere to that process as an important part of the relationship between the parties and a good faith attempt to avoid significant disputes.

Thirdly, if such negotiations fail, then the price revision is usually (but not always) able to be referred to a formal dispute resolution process. In circumstances where there is no provision for a formal process, then there is a serious question as to how any price revision is to be achieved without the agreement of the parties (if at all). Such provisions are often a hangover of an earlier time, where Asia-Pacific buyers and sellers could not contemplate a situation where they would need to resort to formal disputes.[38] Where there is a referral to a formal dispute process, then the forum is usually (but not always) arbitration.[39] Some clauses provided for an alternative to arbitration, with expert determination in particular being popular (which on occasion creates issues between the intersection of that process, with other dispute resolution processes that may apply to other less technical disputes under the SPA).

Fourthly, the clauses often (but again, not universally) set criteria for the price review, including by specifying particular factors to be taken into account.[40] These criteria are often very Asia-Pacific focused; seeking to benchmark any revised price as against prices paid by buyers and sellers within that specific region. This can often lead to detailed disputes as to the precise criteria imposed by the relevant price review clause (ie, what 'market' is relevant to the revision), in addition to the relevant commercial matters (ie, what is the 'market' price). It also means that it is difficult to make generalisations as to the likely outcome of price reviews, as it is heavily dependent on the precise wording of each individual clause.

Finally, although not normally a specific requirement of the price review clause, it appears that all of the price reviews in Asia-Pacific to date have retained the linear formula referable to JCC or crude oil. That is, they have involved amendments to the coefficients in the formula or the introduction of other variants (eg, those suggested above) but retained the formula's basic structure.[41]

5. Historical reasons for the lack of formal price reviews in Asia pacific

As can be seen from the above background, buyers and sellers have been motivated to renegotiate their prices in Asia-Pacific at various times, and these

---

38 Ason (see fn 16), 8.

39 Ason (see fn 16), 6–8; Flower and Liao (see fn 4), 353.

40 Flower and Liao (see fn 4), 353; Ason (see fn 16), 9. Often, they are general in nature and sparing as to the factors (if any) that must be taken into account.

41 Ason (see fn 16), 9. However, note that some have changed, for example, the index used for measuring the crude oil price.

have often involved dealing with challenging problems, sometimes taking years to resolve.[42] Why, then, has there been so little arbitration in relation to this topic? A number of factors have contributed to this.

First, there is a contractual reason. Arbitration is an option only where the parties have made it available. As compared with other regions, it has not been as common for SPAs in this region to provide explicitly for arbitration in the context of pricing disputes.[43] As explained above, before the 1990s Asian SPAs generally did not include price review clauses and, even after they were introduced, they did not always include a direct route to arbitration following failed negotiations. By contrast, in Europe, price review clauses – including provision for international arbitration – became standard in the 1980s.[44]

Secondly, as explained, the relevant clauses in the Asia-Pacific SPAs often specify that price reviews are to be conducted, and can only be conducted, at specific intervals (usually, every five or ten years). The fact is that the bulk of the LNG projects developed in this region have either only just recently, or are now, reaching a level of maturity that such time-based triggers can be reached. Given that many of the large LNG projects commenced production in relatively similar timeframes (particularly in Australia), many of the price review provisions also become effective around the same time.

In other words, it is only comparatively recently that contractual price reviews have been triggered.[45] Whilst it is not impossible to trigger a review (particularly a negotiated review) in the absence of a price review clause, the opportunity to do so is more limited.[46] Thus, the number of reviews – and disputes related to those reviews, capable of being sent to arbitration – has been relatively limited in Asia-Pacific.[47]

Thirdly, there has not been the same impetus for disputes as there has been in other parts of the world, where disputes have been triggered by local market events and changes. For example, in Europe, price review arbitrations have been a relatively common occurrence since 2008 (though they began before this).[48] This was driven by a number of factors specific to the European market, including the liberalisation of the natural gas markets in Europe, which eventually created new benchmarks for pricing gas linked to European hub prices.[49] As a result of these changes, the oil-linked prices in existing SPAs became commercially untenable for many buyers and – where they could not agree a

42 And, in some cases, have resulted in significant changes to the pricing structure (Flower and Liao (see fn 4), 352).
43 Ason (see fn 16), 8.
44 Anway and von Mehren (see fn 10), 5.
45 Ason (see fn 16), 12.
46 Ason (see fn 16), 1.
47 *ibid.*
48 J Baily and R Lidgate, "LNG price reviews: a sign of the times" (2014) 7 *Journal of World Energy Law and Business* 1; Anway and von Mehren (see fn 10); Ason (see fn 16), 11.
49 Stern and Rogers (see fn 8), 16–18; Anway and von Mehren (see fn 10); Stern (see fn 8), 44.

solution with their sellers – they resorted to arbitration.[50] A similar change has not yet occurred in Asia, meaning that buyers have not (at least to date) had the same pressure, or incentive, to recalibrate the pricing structures in their SPAs.[51]

Fourthly, this factor has been accentuated by regulatory environments in certain Asian countries which have 'cushioned' buyers from the effects of price increases. Importantly, buyers have often been able to 'pass through' their costs to their sub-buyers in their home markets, and there has been limited competition as between buyers.[52] In such an environment, buyers have faced less pressure to pursue aggressive dispute strategies in the face of high prices, than they would in a more competitive scenario.

Fifthly, there are reasons related to the preferences of Asian parties over dispute resolution methods. There is a distinct tendency in certain Asian countries, particularly Japan, to resolve disputes via negotiation rather than formal mechanisms (particularly where the primary buyers and sellers share a common nationality, and often related parties form part of the buyers and sellers).[53] There is also a strong perception that pursuing arbitration will be detrimental to ongoing commercial relationships, which militates against the use of arbitration where continued work between the parties is required or desired.[54] In addition, specific concerns have been observed in Asia relating to the outcomes of gas pricing arbitrations in other regions, and their unpredictable nature – for example, the fact that tribunals have materially altered the pricing formula[55] or rendered a decision which diverged from both parties' submissions[56] has given parties reason to pause.

Sixthly, there is potentially a higher degree of uncertainty regarding the outcome of price reviews, if referred to a formal dispute resolution process, as compared to other regions. As explained, in Asia-Pacific there is no ability to refer to common hub pricing or other form of transparent market. Rather, market data has to be sourced from either industry information (which can be, at best,

---

50 These cases often resulted in significant price decreases for the buyer – initially, by changing the level of prices but retaining an oil-linked formula, and eventually the formulae were changed by becoming 'de-linked' from oil. High-profile cases include those between Italian purchaser Edison and RasGas (Qatar) and Sonatrach (Algeria) (awarded in 2012 and 2013). In both of these cases, the tribunal severed the oil price linkage in the contract and awarded Edison substantial sums. Another example is case between RWE, purchaser from the Czech Republic, and Gazprom Export (awarded in 2013), where the tribunal removed the link to oil price and substituted it with a link to gas spot prices. In 2016 an award was made in arbitration proceedings between Edison and Eni (for Libyan gas), which reportedly reduced the price payable by Edison by EUR 1 billion.

51 This is changing, as will be explained below.

52 Stern and Rogers (see fn 8), 17–18. Especially in the traditional Asian LNG importing countries (Japan, Korea and Taiwan).

53 HW Sullivan, "LNG sale and purchase agreements" in P Griffin (ed), *Liquefied Natural Gas: The Law and Business of LNG*, 3rd edn (Globe Business Publishing, 2017) 185.

54 Flower and Liao (see fn 4), 372.

55 Ason (see fn 16), 12.

56 This has been perpetuated by the outcome of a well-known case, *Gas Natural Aprovisionamientos SDG v Atlantic LNG Company of Trinidad and Tobago*, an arbitration commenced in 2005 between Altantic LNG (as seller) and Gas Natural (as buyer), where the tribunal departed materially from the parties' arguments. The case entered the public domain due to being appealed in US courts.

unreliable and speculative) or what can be gleaned from compulsory document production processes compelled by local courts (which can be both very difficult and, in any event, incomplete). Further, the price review clauses are often complicated in terms of establishing the benchmark against which the revised price is to be set, with multiple interpretations both available and difficult to predict. This can be exacerbated where the review is referred to expert determination, rather than arbitration. All of this means that the outcomes of price reviews can be difficult to predict, and do not necessarily follow a consistent trend within the market. Thus, parties often consider that the best way to retain some control over the outcome is to reach a mutual agreement.

6. What next?

It is expected that arbitration will start to play a more central role in resolving gas price disputes in Asia-Pacific. The factors that will likely contribute to this are varied, and there are different views on what will be the major driver of change.[57] It does seem likely, though, that change is on the horizon. A number of reasons support this hypothesis.

Perhaps the primary one of these is practical: as explained above, many of the LNG projects (and associated SPAs) are only now reaching a level of maturity such that clauses providing for price reviews at set intervals are only being engaged for the first time.

Further, there is likely to be heightened impetus for parties in Asia to pursue price reviews and take their disputes to arbitration. Market dynamics are changing in the region, in a way that will impact the viability of the prices as currently formulated in long term SPAs (especially from buyers' perspectives) to a degree that is likely to precipitate disputes and recourse to arbitration.

As foreshadowed above, the Asian market for natural gas is, at present, fundamentally different to the US and European markets, in that it is not fully liberalised. Natural gas and LNG is not freely tradable (around Asia or, usually, even within countries) and there is no 'hub' for trading LNG which works to establish a transparent market price.[58] As such, distinct regional markets have been able to co-exist in the region, quite isolated from one another, and the JCC-linked formula has been able to survive even where it produces prices which are 'out of sync' with other, global, LNG prices. This is in marked contrast to the US and much of Europe, where natural gas prices have been decoupled from oil.[59]

But this situation is coming under pressure. When oil prices are high, many

---

57 Some think that Asia will undergo a period of transition, much like Europe, which will drive the change (Anway and von Mehren (see fn 10)); others think change is more likely to come from a more subtle, region-specific set of factors (Ason (see fn 16), 13).

58 As distinct from Henry Hub in the US and the NBP and TTF (UK/Europe). N Trimble, "Changing LNG markets and contracts" (2018) 11 *Journal of World Energy Law and Business* 432–33.

59 Since the 1980s in the case of USA and late 1990s in UK. In Europe, from late 2000s, as a result of extensive, multi-faceted deregulation efforts: Stern (see fn 8), 45; Stern and Rogers (see fn 8), 1, 15, 16.

Asian buyers have been paying more for their LNG than anywhere else in the world, in a way that has seriously impacted their industrial competitiveness.[60] Thus, there is a general view amongst Asian buyers that something needs to change in the pricing of their long term supplies. Ideally, they want a pricing formula that more accurately reflects the supply and demand conditions in their markets.[61]

To this end a number of Asian buyers[62] and governments of Asian LNG importing countries[63] have advocated a move away from the JCC or crude oil linkage.[64] This has been on the agenda for some years, but the question is: what is the alternative?[65] It will likely take time for a viable alternative to emerge in the region. One option is to establish trading hub or hubs for LNG in Asia, and use the prices generated here as a basis for pricing long term supplies.[66] Various Asian countries – including Singapore, Japan, China and Thailand – are presently attempting to create these in their territories.[67] Each face unique challenges and the general view is that it will take a long time before they will be able to produce a suitable benchmark.[68] There are other options,[69] such as using one of the Asian LNG spot price[70] or other[71] indexes. This is also likely to take time.[72]

If and when a viable alternative is created, this could trigger disputes and arbitrations (similar to the European experience), as buyers seek to reset the

---

60 Oil prices reached $100/bbl in 2011 and have subsequently remained at historic high levels (Stern and Rogers (see fn 8), 12). At prices above $100/bbl for JCC, it generally becomes untenable for buyers (Stern (see fn 8), 46).

61 In other words, more accurately reflective of "market fundamentals": Stern (see fn 8), 47. Note that there is some division of opinion as to whether the JCC link needs to be rehauled, or whether this can be retained but the *level* of prices reduced.

62 From traditional importing countries as well as emerging LNG buying markets in Asia, such as India and Pakistan (Ason (see fn 16), 4). Osaka Gas has been a key proponent (Flower and Liao (see fn 4), 359), as has Japan's largest LNG importer, JERA (Stern (see fn 45), 3).

63 For example, in May 2016 the Japanese Ministry of Economy, Trade and Industry (METI) published the *Strategy for LNG Market Development,* which sets out Japan's intention to be a leader in creating conditions for a global LNG market and the intention for the traditional JCC-linked pricing to be replaced with a new era of market-based pricing: Stern (see fn 45), 1. See also Stern and Rogers (see fn 8), 40. The Pakistan government has recently advised LNG importers in Pakistan to renegotiate their contracts with a view to considering alternative pricing indexations to oil.

64 Stern (see fn 8), 1. Flower and Liao (see fn 4), 338. This has been on the agenda since 2010 and is increasingly being discussed. Note that alternative ways of pricing in spot and short-term contracts have already been developed.

65 GAR, "Are Asia's LNG players heading for a contractual train wreck?" (November 2015), summarising opinions expressed by Professor Jonathan Stern.

66 International Energy Agency (IEA), *Developing a Natural Gas Hub in Asia: obstacles and opportunities* (Paris, OECD/IEA, 2013).

67 ICIS, Global Energy Outlook 2019 Series, LNG Outlook 2019 part one: Asia, see www.icis.com/explore/resources/news/2019/01/10/10304951/lng-outlook-2019-part-one-asia/.

68 Singapore, whilst fairly liberalised, is a small market; Japan and China, while large markets, have a way to go in their liberalisation efforts, before the conditions are met to establish a market price (M Fulwood, "Asian LNG Trading Hubs: Myth or Reality" (Columbia SIPA, Center on Global Energy Policy, 2018) 25–30; Stern (see fn 45), 3).

69 Also, the solution need not necessarily be the same for all countries in Asia: Stern (see fn 8), 48.

70 Such as the Japan Korea Marker ('JKM', an assessment of spot LNG prices published daily by Platts), the Japan/METI (average price of all the LNG imported into Japan each month), the ANAE or SLInG (Fulwood (see fn 67), 33).

71 Such as the RIM Japan or the EAX (Fulwood (see fn 67), 33).

72 That is, before they become sufficiently reliable to be used as a benchmark for long term contracts: (Fulwood (see fn 67), 35). Some buyers in emerging LNG importing countries have also suggested using a 'cost plus' pricing method, which sellers have generally rejected (Flower and Liao (see fn 4), 369).

pricing formula under long-term contracts. More generally, steps are being taken at a country level to deregulate the natural gas markets in Asia. Whilst this will be a slow process for some countries, a commitment appears to exist and some achievements have already been made. As these matters develop, it is increasingly likely to be reflected in the drafting of price review provisions.

Yet even before Asian markets are fully liberalised or an alternative to JCC is presented, there is likely to be pressure on Asian buyers to engage in price reviews.[73] A major factor is that alternative sources of LNG supply are becoming available to Asian buyers, outside long term contracts, including sales under short-term and spot LNG contracts, especially from the US and Australia.[74] These are often priced on a different basis (eg, linked to Henry Hub prices in the case of US imports) and – at least in present market conditions – are considerably cheaper.[75] In addition, as a result of liberalisation efforts, buyers will have less capacity (than previously) to pass on their costs to their customers, and will face more competition between each other.[76] As such, buyers are expected to be more assertive and aggressive in their approach to pricing than they have been in the past, including, where necessary, resorting to arbitration. No longer can they 'sit back' and afford to accept high prices generated by their long-term contracts. Thus, even if there is no option or desire to alter the JCC link, disputes may be on the horizon as buyers seek to reduce the levels of the prices in their long term contracts.

The above issues could, of course, be resolved by negotiation (as has historically been the case), but there is reason to expect that formal dispute resolution will be necessary, due to the complexity of the issues and the heightened pressure parties will be facing to achieve a particular result.[77]

This is particularly so given the sheer scale of the LNG projects that have recently been developed within the region. As explained above, formal disputes regarding price revisions were often avoided, on the basis that buyers and sellers often shared similar nationalities and sought to resolve issues via negotiation. Such an approach has come under significant pressure as parties have formed

---

73 Stern and Rogers (see fn 8).

74 Stern (see fn 8), 44, 47; Trimble (see fn 57), 437; www.smh.com.au/business/global-gas-glut-strengthens-hands-of-buyers-in-pushing-to-slash-prices-20170324-gv5f41.html; International Gas Union, *2019 World LNG Report*, p21.

75 Since the global recession in 2008 there has been a divergence between prices under long-term, oil-indexed contracts and short-term/spot contracts based on hub prices. This is because oil demand and oil prices have generally remained buoyant internationally, whilst gas demand (at least in some markets) has fallen (Bailey and Lidgate (see fn 47), 148).

76 Ason (fn 16), 7; Trimble (see fn 57), 433. New buyers are also emerging in some of the major markets (such as China) as infrastructure is being made available to them.

77 Global Arbitration Review, "Are Asia's LNG players heading for a contractual train wreck?" (November 2015), citing opinion expressed by Professor Jonathan Stern: "It would be nice to think that Asian gas players will find consensus. I hope they do, but I actually think the more likely outcome in the next few years is that we will see a move towards litigation despite the lack of precedents and the relationship culture that we've always traditionally had in Asia." See also Stern and Rogers (see fn 8), 47; Stern (see fn 45), 4.

joint ventures with companies from a broader geographical range and, amongst other things, jointly market LNG. In those circumstances, the ability to reach a consensus comes under great challenge, and just one disaffected party out of a larger number can cause negotiations to fail (and arbitration be required).

There are also other (non-market) factors that suggest an increased role for arbitration in resolving gas price disputes in Asia.

International arbitration is proliferating in Asia generally. It is becoming increasing easy to facilitate, and increasing popular.[78] This will be likely to support its application to LNG pricing disputes, especially as Asian buyers and sellers are already accustomed to using arbitration for non-pricing disputes relating to LNG (eg, infrastructure disputes).[79]

As a result, parties in the Asia-Pacific are increasingly recognising that arbitration is well-suited to resolving these kinds of disputes[80] – or, at least, is the best alternative where formal dispute resolution is required.[81] Some of the key advantages in this context are its confidential nature, the ability to enforce awards internationally and the ability to select tribunal members with the appropriate level and mix of expertise.

There is certainly room for improvement – in particular, by reaching consensus on some of the unique substantive issues that arise in gas pricing disputes.[82] If and when it happens, this will likely enhance the appeal of arbitration for Asian parties. Further, the fact that the first arbitration has been commenced in Asia-Pacific is significant and will, at a minimum, raise awareness in the region that arbitration is a viable option. Coupled with this, Asian buyers may also be inspired by the positive outcomes many European buyers obtained in their arbitrations.[83]

Finally, some of the concerns relating to the unpredictability of arbitration can be overcome by drafting – for example by limiting, via drafting, the tribunal's scope or mandate in terms of the extent to which it can alter the pricing formulae.[84] As parties and practitioners in Asia-Pacific start to consider and commit to the use of arbitration in this context, they will be able to tailor their experiences to meet their preferences.

The author gratefully acknowledges the assistance of Eliza Wallace, Associate, in the preparation of this chapter.

---

78 For example, new case filings at the Singapore International Arbitration Centre have increased by more than four times in the last decade. India, Malaysia, China and Indonesia were key contributors to this caseload in 2018 (SIAC Annual Report, 2018). The case load at the HKIAC has also been steadily increasing, with the top three user countries in 2018 being Hong Kong, Mainland China and Singapore, South Korea and Thailand also within the top 10 (HKIAC Annual Report, 2018).

79 Ason (see fn 16), 12.

80 Ason (see fn 16), 11, 15–19.

81 Global Arbitration Review, "Are Asia's LNG players heading for a contractual train wreck?" (November 2015).

82 *ibid.*

83 van Geuns (see fn 32), 2.

84 Ason (see fn 16), 16–17.

Current and future trends for price reviews

James Freeman
Allen & Overy LLP

1. Introduction

The past decade has seen a boom, in Europe at least, which, in the relatively rarefied world of price reviews and price review arbitrations, has been unprecedented. It has resulted in rigorous and repeated testing of the price review process by parties, their lawyers and economists, and by arbitral tribunals. The purpose of this chapter is not to look back at the boom, which is largely a matter of history rather than current interest, but to ask what the trends are in this field and whether any forecasts can be made about where the market is heading. In particular, the chapter poses three questions:

- What is the trend for price review arbitrations?
- What is the trend for price review clauses?
- What are the market trends under long-term gas and LNG supply contracts?

In considering these questions, this chapter seeks to draw together the threads from some of the earlier chapters by looking back at developments over the five years since the first edition of this book and seeking to identify what the future may hold.

2. The trend for price review arbitrations

It makes sense to start with the most fundamental question for the subject of this book: is there any future for price review arbitration at all in the gas and LNG industry?

2.1 Trends in long-term gas and LNG contracts

The answer to this question can be taken in stages. A first stage is to consider market trends. Price reviews exist because parties recognise that prices in long-term contracts may require periodic adjustment over a 20-year or 25-year term. If long-term contracts were to cease to exist (eg, if gas were sold only under short-term contracts), there would no longer be any need for price reviews. In recent years, there is data suggesting that the average term of new LNG contracts has shortened significantly. For example, analysis by Poten & Partners suggests that the average duration of a new LNG contract fell from around 15 years in 2013 to around seven years in 2017.[1] It seems that a lower proportion of LNG is being sold under long-term contracts. It is more a question for an energy economist as to whether this change is entrenched. It may be a question of negotiating power as some say that it has been a buyer's market in certain recent years. If contracts were to continue to have relatively short durations, price reviews (and any resulting arbitrations) are likely to become rarer. This is because the commercial rationale for a price review is that a price formula agreed in a long-term contract may not reflect market pricing over a term of 20 or 25 years.

In any event, there is evidence that, while more LNG is being sold under shorter-term contracts, new long-term contracts continue to be entered into. This is clear from the same analysis by Poten & Partners, which shows between five and 20 new long-term LNG contracts each year between 2013 and 2017. One reason for this continuing to be the case is that at least some new projects still need evidence of long-term security of offtake to obtain project finance. While long-term gas and LNG contracts exist, there will be a need for price review clauses.

2.2 Trends in price review arbitration clauses

The next stage in the analysis is to consider whether, even if there continue to be price review clauses in long-term contracts, it follows that there is a future for price review arbitration. Historically, long-term contracts, notably in Asia, have tended to contain price review clauses which do not provide for arbitration in the event that the parties fail to agree on a price revision.[2] Depending on their terms, such clauses are likely to amount to nothing more than good faith negotiation clauses, which are generally unenforceable under English law.[3] Such clauses continue to be seen in modern long-term contracts. However, evidence suggests that they may be on the wane;[4] parties seem to be increasingly aware of the potential need to enforce a price revision if necessary.

2.3 Trends in the form of dispute resolution for price reviews

For those contracts which provide for price review arbitration, the next stage is to consider whether arbitration remains the most appropriate form of binding dispute resolution. There are critics of the system. It is argued, for example, that arbitrators are skilled at resolving legal questions of liability and quantum in a contentious proceeding, but that a price review is unlike other contentious proceedings because its purpose is in essence to recalibrate a key commercial

1 Poten & Partners, "Collapse in LNG contract lengths and volumes raises future supply concerns" (*LNG in World Markets*, January 2018).
2 Agnieska Akon, *Price reviews and arbitrations in Asian LNG markets*, OIES paper NG 144 (April 2019) p1.
3 *Walford v Miles* [1992] 2 AC 128.
4 Jonathan Stern, "LNG Pricing: challenges in the late 2010s" in Corbeau and Ledesma (eds), *LNG Markets in Transition: The Great Reconfiguration* (OUP, 2016), p488.

term – the price. It is said to be risky to leave such an inherently commercial exercise in the hands of lawyers.

The first edition of this book contained a chapter on the future of price reviews by the author's late colleague Frances van Eupen and former colleague Paul Griffin. In it, they asked "the question of whether arbitration is the best forum for price reviews", arguing:

> *Under the arbitration process, the LNG contract will be read in the context of an adversarial relationship and a procedure of dispute resolution. This is likely to polarise the parties into self-interested positions, which is the opposite of the consensual nature of both their original bargain and their agreement for any periodic re-setting of the price provisions.*[5]

They went on to discuss, with differing degrees of enthusiasm, various alternatives, including expert determination analogous to its use in unitisation disputes; 'baseball' (or 'pendulum') arbitration; and mediation. It is salutary to consider whether any of those alternatives have gained in popularity.

Certain of these alternatives are occasionally seen in price review clauses. For instance, expert determination is occasionally used instead of arbitration to decide whether, and if so how, the price formula should be adjusted if the parties cannot agree. However, this remains rare, for understandable reasons.

As an initial point, expert determination would not address the ill perceived by van Eupen and Griffin; a typical expert determination may not be as contentious as arbitration, but it remains an adversarial process.

More generally, expert determination can be sensible when there is a purely factual issue on a narrow point such as the certification of completion accounts in an M&A transaction, or whether delivered goods meet certain quality standards. However, pricing disputes rarely concern only narrow factual issues. They usually pose questions of law, notably about how the price review clause should be interpreted, and potentially about wider questions concerning competition law, for example. Moreover, while the factual questions that arise can be 'technical' (eg, issues around the value of gas at a specified date), they can also require a nuanced assessment of evidence. For instance, questions of whether certain events, such as an increase in competition or a change in the supply-demand balance, satisfy the trigger requirements do not lend themselves to straightforward quantitative answers. These factors tend to mean that parties prefer to submit price review disputes to arbitrators, with their experience of fact finding and legal interpretation, rather than to expert determination.

Parties may also have more fundamental concerns about the enforceability of an expert's decision; there is no equivalent to the New York Convention for expert determination.

---

5 Paul Griffin and Frances van Eupen, "The future for price reviews" in Mark Levy (ed), *Gas Price Arbitrations*, 1st edn (Globe Law and Business, 2014), p149.

'Baseball' arbitration is another procedural mechanism which is sometimes seen in price review clauses. It is only a variant on 'traditional' price review arbitration, rather than a completely different approach. Each side must put forward a single proposal for the contract price that should apply from the review date, and the arbitral tribunal must choose one of them. The name derives from its origins in wage negotiations in US baseball. The idea is that this approach provides an incentive for each party to make a reasonable proposal, because otherwise the tribunal will prefer the other side's proposal. It also precludes the tribunal from proposing an adjustment on its own initiative, as appeared to happen in the *Atlantic LNG* case (discussed, for example, in the chapter in this book on the adjustment phase in gas price reviews under long-term gas supply contracts). Commentators have suggested it as potentially a useful mechanism for price review arbitrations.[6]

Baseball arbitration may look good in theory but in practice it can lead to unattractive situations. Any adjustment to the contract price in a long-term gas or LNG contract is likely to be the result of a number of different decisions by the tribunal about (for example) which event has satisfied the trigger requirements, what the end-user price of gas is in different sectors, what is an appropriate gross margin in each sector, and so forth. It is quite likely that the arbitral tribunal does not fully agree with the pricing proposal made by either party because it may agree with one party on certain issues and the other party on certain other issues. It is arguably unattractive that the tribunal should be bound to accept the proposal of the party with which it disagrees less. Baseball arbitration is better suited to situations where there is a simple choice to be made between two competing figures; this is rarely the situation in gas and LNG price reviews.

Moreover, the assumption underlying baseball arbitration is that the parties are incentivised to make reasonable pricing proposals on the basis that their proposal is more likely to be preferred by the tribunal. However, that assumption may be misplaced. For example, in a price review in which the evidence tended to suggest that some degree of hub indexation was appropriate, a party may not be in a position to acknowledge that in its submissions if, for instance, there are strong institutional policies in favour of oil indexation.

Compulsory mediation cannot, of course, constitute a solution for pricing disputes because a resolution depends on the parties' consent. There has to be a binding fall-back mechanism if a settlement is not forthcoming. Even with this in mind, it is rare to see a provision for compulsory mediation in a price review clause. There are grounds to be sceptical about the utility of compulsory mediation at an early stage in any dispute resolution process. If the parties

6 Philip Spencer Ashley and Ben Holland, "Gas Price Reviews: past, present and future" (2011) 6(5) *Global Arbitration Review*.

consider it useful to have a third party to facilitate settlement discussions, they will engage one voluntarily. If they do not consider it useful, a compulsory mediation process is likely to be a waste of time and costs. The same factors apply with perhaps even more force in a price review context, where skilled commercial negotiators, often with a long-standing relationship, are involved on each side. Moreover, most price review clauses provide for a lengthy period of compulsory negotiations of up to six months. Compulsory mediation seems relatively unlikely to generate many settlements after this. In this context, the added value of a mediator is doubtful, although it should not be ruled out in an appropriate case.

Consequently, where the price review clause provides for a binding resolution by a third party in the event that the parties are unable to agree on the outcome of the price review, arbitration (but not baseball arbitration) remains the most popular option in new price review clauses.

2.4 Trends in the practice of price review arbitrations

Nevertheless, while there seems not to have been a move away from arbitration, the experience of price review arbitration over the past few years seems to have resulted in changes in how it is practised.

In particular, it appears that tribunals have, at least in some instances, become more cautious about how they approach a price revision. The author's first price review arbitration, nearly a decade ago, lasted around nine months. The tribunal in that case rendered a single award on all issues, including the appropriate price adjustment, after a short timetable of written submissions and evidence. There remain good commercial arguments as to why these arbitrations should be completed quickly. Both parties wish to know how (if at all) their bargain is to be recalibrated. The "losing" party may be liable to pay a substantial reconciliation payment to reflect the difference between the old contract price and the revised version between the review date and the date of the award, and this will only grow in size as the price review continues. Buyers in particular may find themselves buying gas at a price which leaves them unable to resell it profitably, pending a price adjustment. More generally, an experienced tribunal in an appropriate case may be perfectly capable of rendering a sound single final award without recourse to any of the procedural options below.

However, the trend seems to be towards price review arbitrations getting longer. This trend reflects partly changes in the wider disputes market. Price reviews are often very large disputes and may warrant no stone being left unturned. The trend also reflects the procedure of price review arbitration becoming arguably more sensitive and sophisticated in avoiding the perceived missteps in cases such as *Atlantic LNG*.

Tribunals seem to have become especially careful about the adjustment phase. They rightly approach this phase with caution. It is sometimes suggested that this is because the adjustment phase may require quantitative skills which do not form part of their primary specialism. However, the quantitative analysis for a price adjustment is not usually very complicated, and there are ways of a tribunal avoiding the risk of an error by involving experts (whether appointed by the tribunal or the parties), or the parties themselves, in ways discussed below. It is perhaps more accurate to say that tribunals approach the price adjustment with caution because of the commercial consequences of getting it wrong. Small adjustments can have a significant impact on the returns earned by seller and buyer, and that impact can potentially last for the full remaining term of the contract. Adjustments can also lead to very substantial reconciliation payments. For instance, the award in an arbitration between Gas Natural (now Naturgy) and Sonatrach in August 2010 is reported to have resulted in an order for the former to make a reconciliation payment to the latter of nearly $2 billion.[7]

A number of different approaches to the adjustment phase have been discussed. Perhaps the most limited departure involves the publication by the tribunal of a draft award (including the proposed adjustment), which it circulates to the parties for comment before it is signed. This may reduce the risk of a computational error, but is less likely to obviate the risk of a commercially unsatisfactory adjustment. Moreover, an invitation by the tribunal for comments of a technical nature may descend into attempts by the parties to make further submissions of a more substantive nature.

An alternative approach involves the appointment of an expert by the tribunal. The rationale for this approach is that the arbitral tribunal can draw on the help of an individual with the quantitative and market skills that the arbitrators themselves lack, and thereby reduce the risk of a mistaken approach to the price adjustment (and indeed other aspects of the price review). It is also argued that the independent judgment of a tribunal-appointed expert is more reliable than party-appointed experts, whom the tribunal may regard to some degree as favouring their appointing parties.

However, the appointment by the tribunal of its own expert is not necessarily as beneficial as is sometimes perceived. The tribunal will be better placed to give instructions to its own expert once it is familiar with the issues in dispute. At best, that is likely to be possible following a round of pleadings by the parties. In reality, tribunals may immerse themselves properly in the issues only around the time of the hearing. This may lead to the late appointment of a tribunal-appointed expert. Quite possibly, it may also result in an expert who is less well informed than any experts appointed by the parties. Party-appointed experts are likely to be well informed about the issues

7 Reuters, "Gas Natural pays Sonatrach $1.9 bln to end dispute" (14 June 2011).

in dispute if they have been instructed at an early stage and have discussed those issues with the parties and their counsel.

There are other reasons why the appointment of a tribunal-appointed expert may lead to a lengthier procedure. It is likely that, as a matter of due process, both parties are given an opportunity to make submissions on the expert's work and to ask questions of the expert at a hearing. There may be prior due process controversies about, for example, precisely which evidence is made available to the expert. While not a price review case, the ICSID case of *Hrvatska Elektropriveda v Slovenia*[8] serves as an example. The award shows that the appointment in that case of an expert by the tribunal for the damages phase, after the tribunal had first heard evidence from party-appointed experts, added several years to the arbitration.

Furthermore, a tribunal-appointed expert introduces another opinion into the proceedings. The tribunal may appoint its own expert to provide some reassurance about whether its own views are well founded. It may instead find those views cast into doubt by its expert's own opinion which may differ from the tribunal's previously held view. There is a risk of greater confusion rather than greater clarity.

Tribunal-appointed experts are rare in international arbitration generally. The factors above are likely to explain why they are also rare in price review arbitrations, and unlikely to become more common.

Under a further approach, the tribunal stops short of rendering an award on the precise adjustment to the price formula. Instead, it renders a partial award on all of the differences of principle between the parties. For example, it may rule on whether the trigger has been satisfied, how the end-user price of gas is to be ascertained in each sector, what amount to appropriate deductions for TPA costs and margin in a netback calculation, and so forth. There are two main advantages to this approach. A first advantage is that the tribunal avoids having to engage in any quantitative analysis but instead leaves the parties to agree on the precise adjustment based on the tribunal's rulings on matters of principle. A second benefit is that, should a further phase in the arbitration be necessary, the principles underlying the price adjustment should be largely clear. In this way, the tribunal avoids the risk (often seen in price reviews) of submissions by each party on the adjustment which are based on different sets of assumptions.

This further approach has much to commend it. However, notes of caution are required.

A first note is that this approach is not as divorced from quantitative issues as might be supposed. If, for instance, one party claims that an appropriate margin is 0.30 US$/MMBtu and the other argues for a margin of 1.00 US$/MMBtu, and if the tribunal prefers the first party's submission, the tribunal

8 *Hrvatska Elektropriveda d.d. v Republic of Slovenia*, ICSID Case No. ARB/05/24, Final Award, 17 December 2015.

is effectively making an award of value to the first party which is equivalent to 0.70 US$/MMBtu. Even if a tribunal does not calculate the price adjustment, it must engage with valuation issues in order to render a coherent and commercially sensible award.

Secondly, the approach depends on both parties being willing to collaborate constructively to agree the adjustment following the partial award. This can happen. After all, a price review arbitration is one stage in a long-term commercial relationship between the parties under a long-term contract. However, just as long-term relationships can be dysfunctional in life, so too can they be in commerce; and a hard-fought arbitration may put strain on the relationship and make even an apparently straightforward adjustment difficult for both parties to agree. Sometimes, therefore, a partial award on issues of principle is not the end of the price review. Instead, it is only the first stage, and the bifurcation has the effect of extending the process.

As a result, parties do not always favour this approach, if and when a tribunal canvasses it. Nevertheless, in an appropriate case it can be a sensible way for a tribunal to proceed in incremental steps towards an award for a price adjustment.

3. Trends in price review clauses

It has always been rare for two price review clauses to be alike. The price review clause seems rarely to have been treated as boilerplate; distinct provisions have usually been negotiated so that each clause presents its own nuances. Nevertheless, the building blocks of a price review clause, in European markets at least, have generally been similar; sufficiently similar for Mark Levy to discuss the components of a typical price review clause elsewhere in this book. (In contrast, the substantive and procedural conditions for a price review have probably been more varied in Asian markets.[9])

The typical clause has provided that the trigger for a price review should be a significant and long-lasting change in circumstances in the relevant market which is outside the control of the parties. It has provided that, if the trigger is satisfied, the contract price should be changed to reflect the effect on the value of gas in the relevant market of the significant change in circumstances. The clause might also provide that any adjustment must in any case allow the buyer to competitively (or economically) market the gas, assuming sound marketing practices and an efficient business on the part of the buyer.

A typical price review clause has, therefore, been drafted in broad, and quite brief, terms. The reasons for a broadly drafted clause in a contract with a term of perhaps 20 to 25 years are understandable. It is difficult to predict how gas

---

9 See eg, Agnieska Akon, *Price reviews and arbitrations in Asian LNG markets*, OIES paper NG 144 (April 2019).

markets might change over such a long period, and broad language is likely to be better for ensuring that "unknown unknowns" do not fail to trigger a price review when they should qualify as trigger events. Indeed, some of the changes over the past couple of decades could be said to have vindicated this broad approach. For example, two decades ago, European parties may not have foreseen how far the EU's liberalisation programmes would break down the old model of vertically integrated monopolistic utilities in European gas markets. Broadly drafted price review clauses have ensured that this fundamental change has qualified as a trigger.

Equally, the corollary of a broad and brief price review clause is that it may beg as many questions as it answers. This may happen through both ambiguity and omission. For example, in the "typical" formulation set out above, parties and arbitral tribunals have wrestled with questions of ambiguity such as: what makes a change in circumstances significant; what duration is required for a change to be long-lasting; and how far there needs to be a demonstrable causal link between the trigger and adjustment. Question of omission have also arisen, such as whether an arbitral tribunal has the power to change the indexation of the price formula. It is no surprise that there has been some inconsistency in the way that parties and arbitral tribunals have approached these questions.

The uncertainty has been heightened by two further factors. A first factor concerns the factual questions which commonly arise in price reviews. These proceedings tend to be heavily dependent on expert evidence from economists. Even if it is presented clearly, this evidence can be technical and it can be unpredictable how an arbitral tribunal composed of lawyers will interpret it. Moreover, the issues addressed by this evidence can be subtle. Consider, for example, the question – frequently encountered in price reviews – as to whether the supply-demand balance has changed over the review period. The balance between supply and demand cannot be seen in a gas market. Surplus gas is not stacked up in a stockyard like an excess of widgets. It must be inferred, and the evidence for a trend towards either a tightening or loosening of the market is not always obvious. When the evidence is difficult and technical, the arbitral tribunal's reading of it may be less predictable.

A second factor concerns the confidentiality of the arbitral process. Parties and arbitral tribunals are likely to have wrestled with similar issues in other price reviews. However, public information about price review proceedings is extremely limited. Of the dozens of price review arbitrations in the last decade, only a handful of awards circulate in public.[10] Counsel, experts and arbitrators carry knowledge derived from the price reviews in which they have been

---

10 Awards in ICC Cases 9812 and 13504 in "Extracts from the ICC Arbitral Awards: Price Setting and Price Revision in the Energy Sector" (2009) 20(2) *ICC International Court of Arbitration Bulletin* 69–76 and 93–109; *Atlantic LNG Company of Trinidad & Tobago v Gas Natural Aprovisiamientos SDG SA*, UNCITRAL, Final Award dated 17 January 2008.

involved, but this knowledge has limited value given that the sources from which it is derived are usually confidential. Consequently, the wheel has to be reinvented constantly in price reviews, with no guarantee that it will be reinvented in the same way as it had been previously.

For commercial parties, these uncertainties have introduced an element of risk into the process which in many cases has been regarded as unwelcome. This perception of risk has not been merely a matter of unfounded apprehension. Certain decisions rendered in the past few years have provoked surprise among parties and market commentators. For example, it is unnecessary to comment on whether the decision of the arbitral tribunal in the *Atlantic LNG* case was right, to observe that it was unexpected for the arbitral tribunal to set a new price formula for deliveries to New England. The confidentiality of the price review process is likely to mean that other awards exist which surprised the parties involved without attracting the notice of the wider market.

Parties seem to have responded to the uncertainties resulting from their price review clauses by revising their approach to drafting them. As a general observation, price review clauses have become more detailed and more specific.

In some cases, parties have chosen to amend their price review clauses when settling the previous price review to address the uncertainty that had provoked disagreement in the previous price review. For instance, an issue in a number of price reviews has been how to ascertain the market price of gas in the power sector. The issue presents difficulties because supplies of gas to power producers often take place within a group of companies and there is a lack of information about the prices at which any sales of gas to power producers take place. Certain parties have addressed this difficulty by amending the price review clause to specify precisely how the price of gas in the power sector should be ascertained. Similar amendments could be made to address other potentially contentious issues relating to the interpretation of common provisions in price review clauses.

It appears to be relatively unusual to amend the price review clause once a long-term contract is in force. Presumably, consensual amendments are harder against a backdrop of a price review, especially if it has involved arbitration proceedings. However, there seems to be a trend in the price review provisions in new long-term gas and (more commonly) LNG contracts towards more detailed and specific price review provisions. The apparent purpose behind this trend is to reduce the risk arising from uncertainty as to how parties and arbitral tribunals would apply the short-form and general price review clauses seen previously. At the same time, price review clauses seem to be becoming less alike; parties have different ideas about how to reduce the uncertainty. The mechanisms seen have included:

- Clauses which are more specific about the amendments which an arbitral tribunal has the power to make to the price formula: for example, specifying whether the tribunal has the power to amend the indexation and/or the slope, or whether instead the tribunal may only adjust the P zero.

- Clauses which define more precisely the trigger for a price review: for example, that the difference between the contract price and a particular benchmark must exceed specified bounds.
- Conversely, clauses which define a 'collar' that limits any adjustment in a price review to a percentage change within specified bounds.
- Clauses which specify criteria that the tribunal must take into account when adjusting the price formula.
- Clauses which require the tribunal to issue a draft award in the first instance for the parties' comments.
- Clauses which provide for non-price consequences in the event of specified price trends, such as provisions to excuse the buyer from taking, or the seller from having to make available, specified volumes in certain circumstances.

Interestingly, such changes were proposed by van Eupen and Griffin in the first edition of this book:

> *... [I]n light of the surprising rewriting of the pricing formula in the Atlantic LNG decision, there is much to be said for restricting the ambit of a price review arbitration. Potential restrictions could include limitations on the factors the tribunal should or should not take into account in considering the price review and which aspects of the contract or the price formulae could be subject to change. These could be provided contractually in the price review clause.*[11]

This trend towards more detailed price review provisions is a rational response to the real-life experience of price review arbitration over the past few years. Contractual provisions have become more detailed in response to real-life experience in other contexts (one example being the detailed substantive and procedural conditions applicable to warranty claims in M&A), and it is no surprise to see a similar reaction in the price review context.

4. What next for market trends

The crystal ball of an energy economist rather than a lawyer may be better suited to forming a view on the trends for price reviews in the coming years. Nevertheless, from a non-expert perspective, a number of developments can provisionally be foreseen.

This chapter started by noting the boom in price reviews in Europe over the past decade. It is likely that the peak of that boom has passed.

A major factor underlying the boom was the disconnection between oil prices,

11 Paul Griffin and Frances van Eupen, "The future for price reviews" in Mark Levy (ed), *Gas Price Arbitrations*, 1st edn (Globe Law and Business, 2014).

to which most long-term gas and LNG contracts have historically been indexed, and end-user gas prices, which have increasingly tracked hub prices. Under many contracts, this disconnection has been resolved by switching the indexation from oil prices to hub prices. Certain suppliers to the European market, such as Equinor (formerly Statoil), are known to have agreed to move their contracts to hub indexation.[12] Other contracts are known to have switched from oil to hub indexation because an arbitral tribunal has ordered the switch in an award.[13] Subject to various matters which are discussed below, the potential for price reviews is likely to be reduced, because a contract price indexed to a price at a satisfactory hub is inherently more likely to track the wholesale price for gas in the relevant market than a contract price indexed to the price of oil or oil products.

Moreover, even where contract prices remain indexed to oil, the pressure on buyers under some contracts has been reduced. The disconnection between oil-indexed gas prices and hub prices was generally greater when oil prices were higher. Since the fall in oil prices in late 2014 and early 2015, this disconnection has reduced. Prices under some contracts have even been cheaper for buyers with oil indexation than they would have been with hub indexation. Whereas previously these buyers might have been seeking a change in indexation through the price review process, in some cases they are no longer doing so. (However, the disconnection may, of course, return.)

Nevertheless, there are good reasons to believe that market trends will lead to further price reviews. A number of factors are set out below, in summary form; each of them could be, and indeed has been, the subject of more detailed study.

First, a significant number of European gas and LNG supply contracts retain total or partial oil indexation. Where contracts remain oil-indexed, the seed exists for another price review. Oil indexation is retained for several reasons.

- Certain suppliers continue to prefer to price gas under long-term contracts to oil prices. Even if buyers have preferred to switch to hub indexation, the question of indexation may have been left unresolved, for example, by agreeing to a change in price level but leaving indexation unchanged.
- There remain markets in Europe – notably in the East and South – where hub markets remain under development, and hub indexation has not been widely adopted. Thus, the International Gas Union has assessed that in 2018, while gas prices in North-West Europe were predominantly indexed to hub prices, gas prices in the Mediterranean region remained 56% indexed to oil prices.[14]
- As noted above, oil and hub indexation will not necessarily have led to

12 "Statoil breaks oil-linked gas pricing" (Financial Times, 19 November 2013).
13 See, eg, *Analysis of the arbitral award concerning the price revision proceedings between RWE Supply Trading CZ and Gazprom Export completed*, RWE press release, July 2013.
14 International Gas Union, *Wholesale Gas Price Survey 2019 Edition*, pp57–58.

different contract price levels since the fall in oil prices, and so no switch from one to the other may have taken place for this reason.

Secondly, as the "Natural gas price review arbitrations: issues in adopting hub indexation" and "The potential for price disputes in hub-indexed gas contracts" chapters observe, there may be reasons for price reviews, even after hub indexation has been adopted. A number of issues can be contentious, including: the choice of hub; the choice of hub product (or products) as an index; whether, and if so how, to value any flexibility in the contract; how to value the costs of transporting gas to and from the hub; and how far (if at all) the contract price should be at a premium or a discount to the hub price.

Thirdly, now that (so it appears) the majority of contracts for piped gas into Europe are indexed to hub prices, price reviews seem more likely to arise under LNG contracts. There are particular features of LNG contracts which have the potential to be contentious in a price review context.

One feature concerns the costs of regasifying LNG. This can be a substantial part of any netback from end-user or wholesale gas prices to a price for LNG at the unloading point. At some LNG terminals, the cost of regasifying LNG is negotiated on a case-by-case basis, and no information is publicly available about these costs. A question may arise as to the cost of regasification at these terminals. At other LNG terminals, there may be a regulated tariff for regasifying LNG, but access may also be available to the LNG terminal at a market price. That is, a holder of regasification capacity at the terminal may wish to sell it to another market participant. Alternatively, an LNG terminal which has been unable to sell regasification capacity at the regulated tariff might be willing to sell it at a discount, on the basis that some return on its investment in regasification facilities is better than none. Further questions may arise as to whether the cost of regasification is represented by the regulated tariff or market prices; and also, if capacity is being sold at market prices, how those prices should be assessed.

A further feature concerns any flexibility rights under an LNG contract. A description of the possible features of flexibility under an LNG contract is beyond the scope of this chapter.[15] Suffice it to say that, while the volume flexibility under LNG contracts is generally less than under piped gas contracts, the destination flexibility is generally greater. These destination flexibility rights, which are more likely to be rights of the buyer but can also be available to the seller, can give rise to a number of issues in price reviews.

The best known LNG price review arbitration (because it resulted in the only

15 But see, eg, Steven Finizio, "Destination Restrictions and Diversion Provisions in LNG Sale and Purchase Agreements" in Rowley, Bishop and Kaiser (eds), *The Guide to Energy Arbitrations*, 3rd edn (Global Arbitration Review, 2019).

publicly available award) is the *Atlantic LNG* case. The exercise of destination flexibility by the buyer, Gas Natural, was the main driver of the request by the seller, Atlantic LNG, for a price adjustment. The seller argued that diversions by Gas Natural to New England (instead of Spain) justified an adjustment to the contract price.[16] In the end, the tribunal (somewhat controversially) imposed a new price formula which would apply if Gas Natural diverted a specified percentage of cargoes to the New England market.

Atlantic LNG is an unusual case. Flexibility rights are more likely to be relevant for other reasons. One issue concerns the value of the relevant flexibility rights. The option to take LNG to one or more alternative destinations has a value to the buyer. Equally, any option to the seller to require the buyer to deliver the LNG to another destination has a value to the seller. A question can arise as to whether the value of this optionality should be taken into account when determining the contract price. Could flexibility rights of the buyer justify, for example, a contract price at a premium to a hub price? Conversely, if the seller has the right to require the buyer to take the LNG to a different destination, there may be a question as to whether this would justify a contract price at a discount to a hub price. These are not just issues of principle. If it is decided that the value of this flexibility should be taken into account when determining the contract price, disputes can also arise as to how the optionality should be valued.

A related issue that can arise is whether the value of reloaded LNG is relevant to ascertaining the value of gas in a relevant market. Where LNG is supplied under a contract on DES (ie, delivery ex-ship) terms, title to it passes when the LNG is unloaded, usually at one or more specified LNG terminals. There may be economic sense for the buyer in reloading the LNG on to a ship for re-export to another market. This will only make economic sense if the price difference between the original destination market and the re-export market is greater than the cost of reloading the LNG and transporting it to the re-export market. Put another way, the LNG would have a higher value when re-exported than it would if it was regasified and resold in the original destination market. Some price review clauses provide that the value of gas is to be ascertained by reference to a specified national market (eg, "the value of gas in the Netherlands"). It has been argued (most likely by the seller) that the value of gas in the national market of (in this example) the Netherlands should include the value of any LNG stored at LNG terminals in the Netherlands. Whether this argument has substance is likely to be a matter of contract interpretation.

A fourth reason to consider that price reviews will continue arises from the state of LNG markets in Asia. Negotiated price reviews have long taken place in

16 *Atlantic LNG Company of Trinidad & Tobago v Gas Natural Aprovisiamientos SDG SA*, UNCITRAL, Final Award dated 17 January 2008.

Asian gas markets and a number of factors are likely to drive an increase in price review activity in Asia. These include: the growing role of international players, with a greater willingness to escalate price reviews to arbitration, in Asian markets; a wider inclusion of price review clauses in Asian LNG contracts; an increased acceptance of contentious proceedings in cultures (like Japan) which have historically have evinced a strong preference for consensual resolution of disputes; and the benefit of observing that European price reviews have resulted in meaningful commercial improvements for triggering parties.[17]

17 Agnieska Akon, *Price reviews and arbitrations in Asian LNG markets*, OIES paper NG 144 (April 2019).

About the authors

Luis Agosti
Vice president, Compass Lexecon
lagosti@compasslexecon.com

Luis Agosti is vice president in the International Arbitration Energy Practice of Compass Lexecon. He is an economist specialising in gas, LNG and power and has provided expert advice and written and oral testimony before antitrust, regulatory authorities and tribunals. Over the past few years, Mr Agosti has been involved in over 30 gas and LNG disputes in Italy, France, Spain, Germany, Turkey, Belgium, Poland, Czech Republic, India, Qatar, Israel, China, Korea, Mexico and Peru, on issues related to price reviews, estimation of damages following alleged under-deliveries and undue *force majeure* declarations, interruption of supplies, and changes in delivery points, amongst others.

He started his career in Spain, working as a competition economist in energy cases where he advised companies on the assessment of the competitive impact of major energy mergers in Europe, on the impact of regulation on market functioning and the analysis of alleged abusive conducts such as excessive pricing and price fixing.

He is also co-author of several articles related to the liberalisation of the Spanish energy market, competition in the Spanish power market and the role of renewable energy in the Spanish power market.

Augustin Barrier
Associate, LALIVE
abarrier@lalive.law

Augustin Barrier is an associate in LALIVE's international arbitration team, specialising in commercial and investment treaty arbitration in the energy sector, in particular in the electricity, oil and natural gas markets. He has also been involved in litigation and arbitration proceedings in the commodity trading and shipping industry.

He has acted as counsel in numerous international arbitration proceedings, *ad hoc* or under the major arbitration rules, governed by a range of substantive and procedural laws, both common and civil law, in Europe, Africa and the Middle East. He has in particular assisted clients, buyers and sellers in gas pricing negotiations and arbitrations under supply agreements of natural gas and LNG.

He is a member of several professional associations, including the Young Arbitrators Forum of the ICC (YAF), the Young International Arbitration Group of the LCIA (YIAG), CFA-40, Young ICCA and ASA below 40.

Michelle Glassman Bock
Counsel, Wilmer Cutler Pickering Hale and Dorr LLP
Michelle.Bock@wilmerhale.com

Michelle Glassman Bock is a counsel at Wilmer Cutler Pickering Hale and Dorr LLP in Brussels. Her practice focuses on high-stakes arbitrations governed by both civil and common laws, sited in both civil law and common law jurisdictions. She has acted for clients in arbitration proceedings conducted under the auspices of the major arbitral institutions, as well as in ad hoc proceedings. Ms Bock representations cover a wide range of industries, with a particular focus on energy disputes. She has acted as counsel in numerous gas price review arbitrations under long-term supply contracts. Ms Bock is recognised as a leader in arbitration in the *Global Arbitration Review's Who's Who in International Arbitration Future Leaders* and in the ExpertGuides.

Valentina Bonetti
Senior gas analyst, S&P Global Platts
valentina.bonetti@spglobal.com

Prior to joining S&P Global Platts in 2019, Valentina Bonetti was a senior associate in the London office of The Brattle Group. She has over eleven years of experience in litigation and regulation projects, particularly in the natural gas and LNG industry. She has a master's degree in Economics from Bocconi University.

She has worked on several commercial arbitrations, which have included analyses of long-term natural gas contracts, energy regulation and valuation of energy infrastructure.

Oliver Browne
Partner, Latham & Watkins
oliver.browne@lw.com

Oliver Browne is a partner in the London office of Latham & Watkins. He is chair of the London Litigation & Trial Department and a solicitor-advocate. Mr Browne has been an international arbitration specialist for almost 20 years, and has represented companies and states in proceedings conducted under all the major rules as well as on an *ad hoc* basis. Mr Browne also has considerable experience with complex commercial litigation around the world, and arbitration-related litigation in all levels of the English courts.

Mr Browne was named as one of the "next generation of partners setting the disputes agenda" by *Legal Business* in 2015, listed as one of *Legal Week*'s litigation rising stars 2016 and named a "future leader" in *Who's Who Legal* in 2017, 2018 and 2019. He is ranked in *Chambers & Partners UK 2019* for international arbitration and was named as a leading individual for commercial litigation in *The Legal 500 UK 2019*.

Philip Clifford QC
Partner, Latham & Watkins
philip.clifford.qc@lw.com

Philip Clifford QC is one of London's leading international arbitration Queen's Counsel. He has served as global co-chair of Latham & Watkins' International Arbitration Practice and chair of the London Litigation & Trial Department. He is a United Kingdom representative on the ICC Commission on Arbitration and co-author of a leading book on the LCIA Arbitration Rules. A Fellow of the Chartered Institute of Arbitrators, he also sits as arbitrator, giving him a valuable dual perspective on resolving disputes. He is a member of *The Legal 500* Hall of Fame, having been a leading practitioner for over a decade, and is described as "one of the best in the profession" and "renowned for energy cases". *Chambers* also recognises him as a key figure in international arbitration, with clients describing him as "a pleasure to work with and extremely adept at resolving cross-border disputes".

Rahul Donde
Counsel, Lévy Kaufmann-Kohler
rahul.donde@lk-k.com

A "rising star" (*Legal 500*, 2020) and "future leader" (*Who's Who Legal*, 2020) in international arbitration, Rahul Donde specialises in international commercial and investment arbitration. He has acted as arbitrator, counsel, secretary and assistant of the arbitral tribunal in over 30 international arbitration proceedings under both *ad hoc* and institutional arbitration rules.

Mr Donde has specific experience in disputes relating to natural resources (water, oil, gas, renewable energy), commercial and corporate laws in different jurisdictions. Having worked in Switzerland and India for several years, he has expertise in complex commercial and investment disputes in both common law and civil law jurisdictions.

Mr Donde is empanelled as an arbitrator with all major international arbitral institutions. He sits on the boards of arbitration think tanks and holds managerial positions in several international arbitration organisations. He is regularly invited to speak and write on topics of current interest in energy law and international arbitration.

Steven Finizio
Partner, Wilmer Cutler Pickering Hale and Dorr LLP
Steven.Finizio@wilmerhale.com

Steven Finizio is a partner at Wilmer Cutler Pickering Hale and Dorr LLP. His practice focuses on international arbitration.

Mr Finizio has particular experience with oil and gas, financial services, shareholder, joint venture and M&A issues. He has advised clients regarding disputes under the rules of most leading international arbitration institutions and in *ad hoc* proceedings, and governed by the laws of jurisdictions in Europe, Asia, Africa and the US, as well as under investment treaties. He also serves as an arbitrator.

Mr Finizio's publications include *A Practical Guide to International Commercial Arbitration: Assessment, Planning and Strategy* and 'International Commercial Arbitration' in *The Law of Transnational Business Transactions*, as well as chapters in leading treatises on gas supply and LNG disputes.

Mr Finizio is recognised as a leading international arbitration lawyer by various publications, including *Chambers*, *Legal 500* and *Global Arbitration Review's Who's Who in International Arbitration*.

James Freeman
Partner, Allen & Overy LLP
James.Freeman@AllenOvery.com

James Freeman is a partner in the International Arbitration Group in Allen & Overy's London office. He is also a member of the firm's Energy Group. He has conducted arbitrations under all the major sets of arbitral rules and at various seats, with particular experience of the energy sector. Mr Freeman has substantial expertise in gas and LNG price review arbitrations, in Europe and Asia, having acted for both buyers and sellers over the past decade in numerous cases. He is a solicitor-advocate and routinely appears as an advocate in his cases. He has a Diploma in International Arbitration with distinction from Queen Mary University of London, sits on the Steering Committee of CPR Y-ADR, and is a former co-chair of the Young International Arbitration Group (YIAG) of the LCIA.

Christopher J Goncalves
Chair and managing director of the global energy practice, Berkeley Research Group LLC
cgoncalves@thinkbrg.com

Christopher Goncalves is chair and managing director of BRG Energy's global energy practice. He has 28 years of experience in the energy advisory and financial industries, with a substantial focus on LNG, natural gas, electric generation, and renewable energy. He provides expert analysis and testimony for international energy arbitration and litigation matters, and business advisory services for commercial strategy, contract negotiation, economic and market analysis, regulation, project development and financing, and acquisitions and divestitures. His clients have included law firms, independent and national energy companies, regulators, utilities, developers, investors and banks. His relevant experience spans over 50 countries throughout the Americas, EMEA and Asia.

In recent years, Mr Goncalves has provided LNG commercial strategy for short- and long-term sales pricing, target markets and priority customers for leading NOCs and US LNG exporters.

Mr Goncalves has provided expert services on investment and commercial disputes under ICSID, UNCITRAL, the ICC, ICDR/AAA, the RCJ and the Milan Chamber of Commerce. Over recent years, he has served as the lead LNG and natural gas market and industry expert to energy companies and their legal counsel for over a dozen LNG and natural gas price reviews and/or price review arbitrations in Europe and Asia, as well as many other energy-related commercial and investment arbitration matters. For the last several years, *Who's Who Legal* has recognised Mr Goncalves as a leading energy industry and damages expert.

A native English speaker, Mr Goncalves is also fluent or proficient in Spanish, German and Italian, and has intermediate capabilities in Portuguese and Turkish. He received his MA in international affairs and economics from the Johns Hopkins University School of Advanced International Studies and a BA in international relations from Brown University.

Dan Harris
Principal, The Brattle Group
dan.harris@brattle.com

Dan Harris is a principal in the Rome office of The Brattle Group and has over 20 years of experience in the economics of gas and electricity markets. Prior to joining The Brattle Group in 2002, Mr Harris worked for Shell's upstream oil and gas business in the Netherlands for five years in a variety of roles, including the development of economic models for new oil and gas field developments. He holds degrees in engineering and economics from Imperial College and LSE respectively.

He regularly provides advice in gas contract disputes during both the negotiation and arbitration phases, and major law firms and their clients have retained Mr Harris to act as an expert witness on multiple occasions.

He is listed as one of the world's leading experts in *Who's Who Legal: Arbitration 2019*, is a regular speaker at gas and electricity conferences, and lectures at the Florence School of Regulation.

Nikolas Hübschen
Senior vice president, general law & litigation, Uniper Global Commodities SE
nikolas.huebschen@uniper.energy

Nikolas Hübschen joined E.ON in 2007 after working with an international law firm. He held various positions in the E.ON Group, including in the legal departments of E.ON Ruhrgas and E.ON Italia. At E.ON Ruhrgas he dealt with a wide variety of gas-related matters. In 2013 he became Head of Gas Purchase Law of E.ON Global Commodities SE, in which position he in particular managed E.ON's gas price arbitrations. Following the spin-off of Uniper from E.ON in 2016, he is now senior vice president, general law & litigation, of Uniper's Legal & Compliance Department, responsible *inter alia* for the legal aspects of Uniper's gas purchase, gas storage and energy wholesale businesses, as well as the Group's arbitration and litigation proceedings. Mr Hübschen speaks regularly at dispute resolution and energy conferences.

Mr Hübschen holds a diploma and a doctoral degree in law from the University of Münster (Germany) and a Master of Laws from Boston University (USA).

David Jenaway
Partner, Allen & Overy
david.jenaway@AllenOvery.com

David Jenaway is a partner in the Perth office of Allen & Overy. He has extensive experience in a wide range of high profile commercial disputes, with a particular emphasis on matters arising in relation to the construction and energy and resources sectors. He regularly represents large corporations in high value superior court litigation and international commercial arbitration.

Mr Jenaway has experience on a number of LNG price reviews that have been resolved by formal dispute resolution in the Asia-Pacific region, as well as arbitrations that have been conducted as an adjunct to, or in support of, those reviews. He also has extensive experience with price reviews for domestic gas contracts and other similar commodities within Australia.

Mr Jenaway is valued by clients and has been recognised in *Legal 500* and *Chambers Global and Asia-Pacific*. He is listed in the Best Lawyer List for international arbitration and litigation in Australia.

Carlos Lapuerta
Principal, The Brattle Group
carlos.lapuerta@brattle.com

Carlos Lapuerta is a principal and director of London office of The Brattle Group and has been an expert in the energy sector for 30 years. He holds degrees in economics and law from Harvard University.

He has experience in the analysis of investments and contracts in the energy sector, and has provided testimony in several arbitration proceedings concerning the prices in long-term energy contracts, including gas and LNG contracts. He also offers extensive expertise in financial valuation, which he frequently applies to estimate damages in international arbitration proceedings, and in the analysis of competition.

He speaks regularly at energy and arbitration conferences, and lectures at the Florence School of Regulation. *Who's Who*

Legal: Arbitration 2019 lists Mr Lapuerta as a Global Thought Leader.

Marnix Leijten

Partner, De Brauw Blackstone Westbroek NV
marnix.leijten@debrauw.com

Marnix Leijten is a noted international arbitration specialist. His experience includes a wide variety of high stakes international disputes, commercial and investor state arbitrations, as well as complex enforcement and setting aside disputes.

Mr Leijten handles international arbitrations in a wide variety of industry sectors. He has particular experience in advising large corporates in the energy industry and is widely recognised as a leading expert on disputes regarding natural gas and LNG (including supply, transport, trading and construction). He has represented clients on dozens of gas price review disputes over the last 18 years, for a variety of leading gas suppliers and buyers. He also frequently acts as arbitrator, including on gas and LNG pricing disputes.

Mr Leijten was the Dutch member of the ICC International Court of Arbitration from 2006–15, and has been vice president of the Court since 2015. He co-chairs the ICC Commission's Task Force on Emergency Arbitration and is an active member of the ICC Institute of World Business Law, the ICC Commission on Arbitration, ICCA and the IBA Arbitration Committee.

Laurent Lévy

Founding partner, Lévy Kaufmann-Kohler
Laurent.Levy@lk-k.com

Laurent Lévy is a partner and co-founder of Lévy Kaufmann-Kohler in Geneva. He has extensive experience in corporate disputes, including in the areas of oil and gas, air and space and finance industries.

Mr Lévy is a Council member of the ICC Institute of World Business Law, a member of the Association Suisse de l'Arbitrage (ASA) and of the Milan Club of Arbitrators. He is also a former vice president of the ICC Court of International Arbitration and a former member and vice president of the LCIA. He is president of the Casablanca Court of Arbitration.

Mr Lévy has published extensively on international arbitration and is, since 2007, a visiting professor at the School of Law, Queen Mary University of London.

He has handled more than 250 arbitration proceedings, mostly as an arbitrator, under various rules such as the ICC, ICSID and LCIA rules.

Mark Levy QC

Partner, Allen & Overy LLP
Mark.Levy@AllenOvery.com

Mark Levy QC is co-head of Allen & Overy's international arbitration group. He is based in the London office. He has a degree in English Literature from Oxford University and has worked exclusively on international arbitration matters for over 10 years. Mr Levy specialises in energy and commodity disputes. He has extensive experience of acting as counsel in both commercial and investment treaty arbitrations, often as lead counsel. His recent experience has been focused on the oil and gas sector, and in particular liquefied natural gas disputes.

He has acted in some of the largest recent gas price review disputes in the world, involving various markets in Europe as well as markets in Asia and the Middle East. He has also acted in pricing disputes involving commodities such as alumina.

Anthony J Melling
Director, Berkeley Research Group LLC
amelling@thinkbrg.com

Focusing on natural gas, LNG and associated liquids, Anthony Melling has over 20 years of hydrocarbons industry commercial experience with leading international E&P and energy companies, and 18 years of international consultancy experience. As a gas industry professional, Mr Melling offers extensive knowledge of the international gas chain, including upstream, midstream and downstream experience. His experience spans Europe, the Americas, Eurasia, Africa and Asia.

Mr Melling specialises in international natural gas, LNG and liquids pricing, market dynamics and industry economics. In recent years, he has applied these skills in a variety of international energy dispute matters and business advisory engagements. Mr Melling is accomplished in providing these services to law firms, global energy companies and utilities, national energy companies, state energy agencies, independent developers, equity investors and banks. He has advised clients on large-scale energy infrastructure and commerce, including LNG, natural gas, LPGs, oil, conventional power generation and renewable energy.

In late 2010 his book *Natural Gas Pricing and its Future – Europe as the Battleground* was published by the Carnegie Endowment for International Peace.

Boaz Moselle
Executive vice president, Compass Lexecon
bmoselle@compasslexecon.com

Boaz Moselle is an executive vice president at Compass Lexecon based in London. He is an economist who has worked in academia, consulting and government. His areas of expertise include oil and gas, electricity, renewable energy, regulated infrastructure, gas pricing, valuation and the assessment of damages, and regulatory and antitrust issues in the energy industry.

Dr Moselle has provided expert witness testimony in approximately 50 international arbitrations, in the context of both commercial and investment treaty disputes. *Who's Who Legal 2017* notes that Dr Moselle "draws praise for his 'exceptional genius' and prowess in gas-pricing disputes" and that "one client describes him as 'the best expert I have ever seen under cross-examination'".

He holds a PhD in Economics from Harvard and an MA and PhD in Mathematics from the universities of Cambridge and London. Previously a managing director of the UK energy regulator Ofgem, Dr Moselle teaches in the Brussels School of Competition on competition and regulation. He was most recently the Agmon Rosenberg Hacohen & Co Visiting Chair in Business Law at Tel Aviv University, where he taught on issues related to valuation, excessive pricing and the assessment of damages.

Philippe Pinsolle
Partner, Quinn Emanuel Urquhart & Sullivan LLP
philippepinsolle@quinnemanuel.com

Philippe Pinsolle is the head of international arbitration for continental Europe. He has acted as counsel in more than 250 international arbitrations, with a particular focus on investor–state arbitrations and commercial disputes involving the energy, power, oil and gas, construction and defence industries. He has been involved in arbitrations under the auspices of virtually all major arbitration institutions including the ICC, LCIA, ICSID, SCC, AAA, ICDR, the Swiss Chambers of Commerce and the AFA, as well as in *ad hoc* cases under the UNCITRAL rules or otherwise. Mr Pinsolle has also served as arbitrator in more than 50 cases, as well as expert witness on arbitration and French law issues.

In 2019, he was appointed as a member of the SIAC Court of Arbitration. In 2020, he will co-chair the IBA arbitration committee.

Michael Polkinghorne
Partner, White & Case LLP
mpolkinghorne@whitecase.com

Michael Polkinghorne is a dual-qualified (civil and common law) lawyer who heads the White & Case LLP Paris office's arbitration, energy and pro bono groups; he is also joint head of the firm's oil and gas practice. He has served as counsel in arbitrations conducted under most major institutional rules, and has been involved in a number of mediations, notably in the industrial gas sector. He has served in over two dozen cases as arbitrator under, among others, the ICC, LCIA and UNCITRAL rules. Mr Polkinghorne is the former Australian member of the International Court of Arbitration of the ICC, and a current member of the Legal Advisory Task Force of the European Energy Charter Secretariat. Having written extensively on energy dispute resolution and foreign investment, he also provides courses on long-term energy contracts and dispute resolution. He is a graduate of Adelaide (LLB Hons: Stow medalist) and Cambridge (LLM: Commonwealth Scholar) universities.

Noradèle Radjai
Partner, LALIVE
nradjai@lalive.law

Noradèle Radjai is a partner in LALIVE'S international arbitration team, specialising in commercial and investment arbitration.

Ms Radjai has acted as counsel and arbitrator in various international arbitration proceedings, both *ad hoc* and institutional, governed by a variety of procedural and substantive laws, both common and civil law. She focuses on energy matters, including oil and gas. She has acted as counsel and advocate in numerous gas pricing arbitrations, since 2005, for both buyers and sellers, covering the European, Middle Eastern and Asian markets. She has also spoken at several conferences on the topic, including the C5 conference on Long-Term Gas Supply Contracts.

She is a member of the IBA Arbitration Committee, as well as a member of the ASA Board and CPR European Advisory Board. She is also a member of the ICCA, LCIA, ABA and Arbitral Women. Ms Radjai is a solicitor-advocate of the Senior Courts of England and Wales and is fluent in English, French and Arabic.

Martje Verhoeven-de Vries Lentsch
Partner, De Brauw Blackstone Westbroek NV
martje.verhoeven@debrauw.com

Martje Verhoeven-de Vries Lentsch is a partner in the litigation and arbitration department of De Brauw Blackstone Westbroek in Amsterdam. She specialises in commercial arbitration and complex cross-border litigation.

Ms Verhoeven-de Vries Lentsch's particular expertise is in the field of energy, where she advises clients and litigates on a wide variety of energy-related topics. In particular, she has handled and advised on over ten gas price review arbitrations related to long-term, high-value natural gas supply agreements governed by a variety of legal systems, representing both buyers and sellers. This experience combined with her secondment to Shell's litigation department in Houston and her role as litigator at the Houston office of Haynes & Boone enables her to advise clients on both legal and economic issues related to price reviews.

She also regularly advises clients on post-M&A disputes and issues related to recognition and enforcement of foreign judgments and related attachment proceedings.

Index

About Globe Law and Business

Globe Law and Business was established in 2005, and from the very beginning, we set out to create law books which are sufficiently high level to be of real use to the experienced professional, yet still accessible and easy to navigate. Most of our authors are drawn from Magic Circle and other top commercial firms, both in the UK and internationally.

Our titles are carefully produced, with the utmost attention paid to editorial, design and production processes. We hope this results in high-quality books which are easy to read, and a pleasure to own. All our new books are also available as ebooks, which are compatible with most desktop, laptop and tablet devices.

We have recently expanded our portfolio to include a new range of journals and Special Reports, available both digitally and in hard copy format, and produced to the same high standards as our books.

We'd very much like to hear from you with your thoughts and ideas for improving what we offer. Please do feel free to email me at sian@globelawandbusiness.com with your views.

Sian O'Neill
Managing director
Globe Law and Business

www.globelawandbusiness.com

Globe Law and Business
Decarbonising Energy
The Pathway to Net Zero
Hugo Lidbetter
Energy Insights

www.ingramcontent.com/pod-product-compliance
Ingram Content Group UK Ltd.
Pitfield, Milton Keynes, MK11 3LW, UK
UKHW062150170926
8656IPUK00013B/42

9 781787 421929